For [redacted] — with respect
for his [illegible]
significant contribution
to human welfare.

Sincerely,

[signature]

6/21/72

Impact of
New Technologies
on the
Arms Race

Proceedings of the 10th Pugwash Symposium,
held at Wingspread, Racine, Wisconsin, June 26–29, 1970

Sponsored by the Committee on Pugwash Conferences on Science and World Affairs of the American Academy of Arts and Sciences

with the support of
The Johnson Foundation
The Alfred P. Sloan Foundation

Organizing Committee:
F. A. Long and
G. W. Rathjens, co-chairmen
B. T. Feld, S. Weinberg

Rapporteurs:
T. Greenwood and
J. J. MacKenzie

The MIT Press
Cambridge, Massachusetts,
and London, England

Impact of New Technologies on the Arms Race

A Pugwash Monograph

B. T. Feld,
T. Greenwood,
G. W. Rathjens,
S. Weinberg,
Editors

F. A. Long,
J. J. MacKenzie,
Associate Editors

Set in Monotype Univers by
Wolf Composition Co. Inc.
Printed and bound in the
United States of America by
The Colonial Press Inc.

ISBN 0 262 06042 6
(hardcover)
ISBN 0 262 56010 0
(paperback)

Library of Congress catalog
card number: 72–148970

3 Military Research and Development

4 The Political Implications

Preface

The Pugwash Conferences on Science and World Affairs have been held periodically (annually in recent years) since their inception in 1957. They came about in response to a public appeal by Bertrand Russell, Albert Einstein, and nine other distinguished scientists to the worldwide scientific community on the urgent necessity of concerted action to avert the dangers of nuclear war. Since the first Conference, held in Pugwash, Nova Scotia (whence the name), there have been twenty such meetings among scientists and scholars from all parts of the world. Limited at first to discussions of the nuclear menace and means of averting it, the subject matter has been broadened to include all aspects of arms control and disarmament and scientific and technical cooperation in many spheres, including problems of the environment and developing-nations problems.

At the Seventeenth International Conference, held in Ronneby, Sweden, in September 1967, the conferees proposed the organization of a series of smaller, topical symposia having two main objectives: first, to make possible a more professional and concentrated examination in depth by qualified experts of the variety of issues which were and continue to be the concern of the scientists meeting at the annual, broadly representative International Pugwash Conferences; second, as a means of bringing into the Pugwash "movement" a much larger representation of the younger generation of scientists and to provide, through these Symposia, an outlet for the constructive manifestation of their profound concern for the application of their scientific efforts toward the solution of the critical problems of human survival.

This symposium on the "Impacts of New Technologies on the Arms Race" is the tenth in the series of International Pugwash Symposia.* It has been organized by the Committee on Pugwash Conferences on Science and World Affairs (P-COSWA) of the American Academy of Arts and Sciences with the active participation of the National Academy of Sciences. The organizing committee consisted of F. A. Long and G. W. Rathjens, cochairmen, and B. T. Feld and S. Weinberg. The Conference was made possible by the hospitality of the Johnson Foundation, which provided its magnificent headquarters at Wingspread in Racine, Wisconsin, and has supported the expenses of publication of this volume of Proceedings. Our indebtedness to Leslie Paffrath, President of the Johnson Foundation, to his able assistant, Mrs. Rita Goodman, and to their very competent staff is difficult to express in words. Funds for organization and travel were provided by grants from the Alfred P. Sloan Foundation, the Ford Foundation, and the Christopher Reynolds Foundation, while the administrative and logistic support were provided by the staff of the American Academy of Arts and Sciences through its Executive Officer, Mr. John Voss. Whatever success we have achieved would have been impossible without the wholehearted efforts of all of the aforementioned.

* A list of the previous symposia in the series is appended.

From its conception, it has been our idea that this Symposium should serve as a prototype: we have tried to attract the most qualified technical experts from both the so-called defense and the arms-control communities; we have endeavored to elicit original and provocative papers on the issues under consideration; and we have hoped that the atmosphere of the discussions would be open, free, and conducive to new ideas and approaches. We cannot claim complete success in achieving all of these goals. But we believe that the results, as set forth in this volume, justify the efforts of all those who worked so hard to make this Symposium both interesting and productive.

In the correspondence that the organizing committee had before the conference with many of the people who had been invited, we were continually asked the question, "Why have this kind of symposium?" The unusual thing is not that this conference was a symposium on the arms race, but rather that it had such a heavy element of technical content, as will be evident from the papers which follow. The hope of the organizers on planning such a conference was threefold.

First, and certainly most important, when we talk about the future of the arms race we must recognize the important role played by developments in new technology, which, as shown by the curves in Frank Long's paper, have a life of their own. There is a steady upward rise in the spending for research and development which seems to have little to do with outside stimuli. Usually this speculation on future technological developments is done in a rather generalized way. In this conference we were hoping, by bringing more specifically technical elements into the discussion, to look into the seeds of time and see which developments would cause us trouble in the coming years and which, perhaps, might offer some hope. It is not at all clear that we have been able to do this. The first part of our hopes will have been fulfilled if some of the conference participants, or some of those who read these proceedings, are sparked into a realization of some danger or some opportunity provided by a specific element of new technology that was discussed at the symposium.

Second, we gathered together at the symposium people who did not know each other very well. There were people, in particular some of those who prepared key papers, who had not previously been actively concerned with the problems of arms control, but whose center of interest had been in the technology itself. We did not know what would come out of this kind of mixture of interests and expertise, but we felt that the experiment of bringing together such technical experts with members of the arms-control community was worth trying.

The third thing we hoped to accomplish was to put together this volume of symposium proceedings. The participants in the ABM debate last year in the United States were helped tremendously by the fact that there was at least one first-rate technical article, that by Bethe and Garwin in the *Scientific American*, March 1968, on ABM technology, which provided a source of unclassified, open, available material upon which they could draw to make sensible and informed judgments. That article set a very high standard. We feel that some of the contributions to our symposium were of that high standard, and that, therefore, this volume may play a similar role in future discussions on the direction of the arms race and its limitations.

It is necessary to add a word about the method we have used to report the discussions which followed the presentation of papers. We have summarized parts of these discussions, and have for the most part not attributed remarks to individual participants, because it would have been very difficult and time-consuming to get the approval of each participant of our version of his remarks. However, it should be kept in mind that these are summaries of remarks made by individuals who were often in disagreement with each other. No attempt was made to reach a consensus or to report the opinion of the majority, and certainly the discussion reports are not meant to represent views of the editors.

Pugwash Symposia

1	
Control of the Peaceful Uses of Atomic Energy with Particular Reference to Nonproliferation	11–16 April 1968 London, England
2	
Scientific and Technical Cooperation in Europe as a Contribution to European Security	Part I 13–18 May 1968 Marianske Lazne, Czechoslovakia Part II 8–10 September 1968 Nice, France
3	
Implications of the Deployment of ABM Systems	14–20 July 1968 Krogerup, Denmark
4	
Economic Aspects of Energy Production (with Particular Reference to Nuclear Power)	12–17 April 1969
5	
Role of Science and Scientists in National and World Affairs	19–24 May 1969 Marianske Lazne, Czechoslovakia

6

An International Agency for the Collection and Dissemination of Information on Potential Crises

8–12 September 1969
Elsinore, Denmark

7

Arms Control and Disarmament Measures in Europe

9–12 December 1969
Rdziejowice, Poland

8

Overcoming Protein Malnutrition in Developing Countries

19–23 May 1970
Oberursel, Federal German Republic

9

The Setting up of Institutions for European Scientific and Technical Cooperation

5–8 June 1970
Noordwijk, Netherlands

10

Impact of New Technologies on the Arms Race

26–30 June 1970
Racine, Wisconsin, United States

Program

Friday, 26 June	**Introductory Session** Chairman: Franklin A. Long
Saturday, 27 June	**New Technology** Chairman: Steven Weinberg
Sunday, 28 June	**Military Systems** Chairman: George W. Rathjens
Monday, 29 June	**Technology and Policy Planning** Chairman: Bernard T. Feld

Participants

Horst Afheldt
Vereinigung Deutscher Wissenschaftler
Federal Republic of Germany

Ben Alexander
General Research Corporation
United States

Victor C. Anderson
Scripps Institute of Oceanography
United States

Bruno Brunelli
Laboratorio di Gassi Ionizatti
CNEN, Frascati
Italy

George Bunn
University of Wisconsin
United States

Abram Chayes
Harvard University
United States

Paul Doty
Harvard University
United States

Vasiliy Emelyanov
Disarmament Commission, Academy of Sciences
Soviet Union

Bernard T. Feld
Massachusetts Institute of Technology
United States

Shalheveth Freier
The Weizmann Institute of Science
Israel

Eugene Fubini
Consultant
United States

Jules Gueron
Faculty of Science, University of Paris
France

W. Häfele
University of Karlsruhe
Federal Republic of Germany

David G. Hoag
Massachusetts Institute of Technology
United States

G. B. Kistiakowsky
Harvard University
United States

Oleg I. Kozinets
Physical Institute, Academy of Science
Soviet Union

Franklin Long
Cornell University
United States

J. Carson Mark
Los Alamos Scientific Laboratory
United States

Igor Mikhaltsev
Institute of Oceanology
Soviet Union

P. L. Ølgaard
Atomic Energy Commission Research Establishment
Denmark

Jan Prawitz
National Research Institute of Defense
Sweden

B. T. Price
Ministry of Transport
England

R. Ramana
Bhabha Atomic Research Centre
India

George W. Rathjens
Massachusetts Institute of Technology
United States

James Read
The Kettering Foundation
United States

J. P. Ruina
Massachusetts Institute of Technology
United States

Roald Sagdeev
Institute of Nuclear Physics
Soviet Union

Ian Smart
Institute for Strategic Studies
England

Vyacheslav Sychev
Institute of High Temperature
Soviet Union

Robert L. Sproull
University of Rochester
United States

Steven Weinberg
Massachusetts Institute of Technology
United States

Observers

Thomas McCorkle
The MIT Press
Massachusetts Institute of Technology
United States

Vladimir Petrovsky
Department of Political Affairs
United Nations

Rapporteurs

Ted Greenwood
Massachusetts Institute of Technology
United States

James J. MacKenzie
Massachusetts Institute of Technology and Massachusetts Audubon Society
United States

Introduction: Technology and the Arms Race—Where We Stand

G. W. Rathjens

When we were putting together the program for this symposium, we asked Herbert York if he would speak on the subject of where we now stand in the arms race and the role of technology and technologists in it. He would have been an ideal choice both because of his past experience and because he has just finished a book, *Race to Oblivion*, in which he has reviewed the history of the development of major weapons systems and tried to draw some lessons for the future. Although he accepted our invitation, later developments unfortunately made it impossible for him to be here; so it has fallen to me to try to fill in. I decided that in doing so it might be useful to try to tap Herb York's knowledge and wisdom by drawing on his about-to-be-published book.

York makes a number of interesting, and in some cases I believe contentious, observations. I will make others that may be even more debatable, but that I will state without much elaboration or defense unless pressed about them. I should say that I find little to disagree with in what York says but others here may find more. Let me open by identifying and discussing briefly two or three of his observations so that we may have them in mind as we proceed.

George Rathjens was formerly Chief Scientist and Deputy Director of the Advanced Research Projects Agency, United States Department of Defense, Special Assistant to the Director of the United States Arms Control and Disarmament Agency, and Director of the Systems Evaluation Division of the Institute for Defense Analysis. He is now Professor of Political Science at M.I.T.

Perhaps at the conclusion of our meeting we may want to return to some of his propositions for further discussion. York limits himself to the Soviet-American strategic-arms race and so shall I. Much of what we both have to say is however, I believe, more generally applicable.

The overriding theme of York's book is that the strategic-arms race is absurd in two senses.

The first absurdity is in the fact that while the military power of the United States has been increasing steadily since World War II, our security has been diminishing; and the same thing has been happening in the Soviet Union. Few would dispute this. At the end of World War II it would have taken a massive effort by either superpower to inflict more than a few million fatalities on the other. Now well over a hundred million people in each nation could be destroyed in a matter of hours or less.

York's second absurdity is, as he points out, "in an early stage, and for reasons of secrecy is not yet so widely recognized as the first. It lies in the fact that in the U.S. the power to decide whether or not doomsday has arrived is in the process of passing from statesmen and politicians to lower level officers and technicians and, eventually, to machines. Presumably the same thing is happening in the Soviet Union." York attributes his second absurdity, which he calls the ultimate one, "to two root causes. One of these is the development and deployment of weapons systems designed in such a way as to require complex decisions to be made in an extremely short time. The other is in the sheer size and wide dispersal of our nuclear weapons arsenal." I would add a third cause, and that is in the concentration of destructive power in a single weapon and the almost inevitable concentration of power in the hands of very small numbers of men not at the apex of our political or military structure but rather buried far down in it. Thus, we now have in a single Polaris submarine, commanded normally by a medium-level officer, with the capability to deliver 16 nuclear warheads, a number that will shortly be multiplied by a factor of around 10.

The development of these two absurdities is, as York has pointed out, "a manifestation of the interaction of technological developments with the chronic confrontation between the superpowers." I would add that the future is made even more ominous by the prospect of proliferation of destructive power to more and more nations at least some of which may be less willing or less able than the United States and the USSR to limit effectively the authority of individuals or small groups to initiate nuclear war; that is, to cope with York's second absurdity.

At issue before us is the question of whether we can reverse the trends of the last decade by limiting some applications of new technology and perhaps by using technology in the cause of arms control and disarmament.

Each of York's absurdities involves questions of stability of the present strategic balance, and in some sense there is a one-to-one correspondence between his two absurdities and what many of us have come to refer to as two different kinds of stability: arms race stability and crisis stability, respectively.

In referring to the former, one usually has in mind what we commonly call the action/reaction phenomenon of the arms race: the fact that each side tends to react to weapons development and procurement decisions of the other (or even to the possibility of such decisions), with escalation in the arms race being the result. Some technological developments and some decisions regarding weapons may be more likely to produce such escalation than others, and hence would be referred to as more destabilizing. The same technologies or decisions may or may not make the initiation of nuclear war (or its escalation) more likely in a time of crisis.

York identifies four technological factors that threaten to upset the present Soviet-American strategic balance in one sense or the other or in both. These are (1) improvements in reliability, (2) further improvements in guidance accuracy, (3) multiple independently targetable reentry vehicles (MIRVs), and (4) anti-ballistic-missile (ABM) defenses. But he also argues that the present balance is in fact an extremely stable one. It is not delicate in the sense that Albert Wohlsletter characterized it a few years ago but rather is very resistant to militarily significant perturbation.

In the remainder of my remarks I would like to explore further the question of whether the present balance is in fact stable in both of the senses to which I have referred, and whether or not it is likely to be upset by technological developments during the next few years. I will also attempt to reconcile the contention that it is improbable that new systems developments or procurement will be militarily or politically useful with the contention that in the absence of effective arms control measures we will develop and buy a number of new strategic systems over the next decade.

Assuming no effective counteraction by the Soviet Union, U.S. intercontinental bombers and long-range missiles could now deliver some 4,000 nuclear weapons against the Soviet Union, each with a yield of a megaton or more.* Delivery of one-tenth of this number would suffice to destroy three-quarters of the Soviet industrial capacity and over one-third of the population. Delivery of one hundred would destroy 60% of the industry and over 15% of the population.

While the number of deliverable Soviet weapons is smaller, their yields tend to be larger. Moreover, the United States may be a more vulnerable country than the Soviet Union because of a higher degree of urbanization. Therefore, to a first approximation the situation is one of strategic parity.

Under almost any conceivable circumstances a very large fraction of either nation's strategic weapons would be deliverable. Thus each nation now has a strategic force that, in the absence of crisis, is

* Each Polaris A-3 missile carries multiple warheads of smaller yield, but the effect of each payload is expected to be about the same as would be achieved were each missile to carry a single warhead in the megaton range. The figure 4,000 is based on counting each A-3 payload as a single warhead.

clearly many times larger than required for deterrence of any attack by its superpower adversary that might reasonably be expected to induce retaliation.

But one might ask about the validity of the same conclusion in a crisis situation. If a nuclear war seemed imminent and probable, would not there be powerful incentive to launch a massive attack, not in the hope of escaping a devastating retaliatory blow, but rather in the hope of limiting the severity of damage that would be inflicted in a war that seemed inevitable anyway? While the incentive might be great, the actual initiation of a preemptive attack would seem unlikely in the present context because of the enormous degree of overkill capability on both sides. Even if there were an expectation of being able to knock out a very large fraction of the adversary's strategic force—perhaps all of one of its components, for example the ICBMs—such an attack would still seem totally irrational. It could not be justified by the savings in lives and property, which would be small, as long as there were any chance whatever of avoiding a holocaust.

This is the basis for contending that the present balance is very stable in the crisis-stability sense. But clearly the present situation is not one of stability in the arms race sense. The rapid growth in Soviet strategic forces and recent American decisions to introduce major changes in our strategic forces—the introduction of MIRVs, the proposal to defend our Minutemen bases with an ABM system, the decision to go ahead with development of the B-1 bomber and the initiation of work on other new strategic systems such as the underwater long-range missile systems (ULMS)—all illustrate that.

The seeds of difficulty are, at least on the American side, in what we have come to call "worst-case analysis." The present situation is one of extreme instability in the arms-race sense not because it is actually *likely* that implementation of any of the development or procurement options before us would in fact alter the military balance in ways that could be exploited militarily or politically, but rather because they just conceivably *might*. In a situation in which there are large uncertainties about the capabilities of one or both sides and perhaps also large uncertainties about adversary intentions, it is common in analysis to give one's adversary the benefit of all doubts and one's self the benefit of none, i.e., to assume the worst with respect to adversary intentions and that his military systems will perform at the upper limits of what seems technically feasible while one's own systems perform at lower limits of the range of uncertainty. The tragedy of such an approach, which we refer to as worst-case analysis, is that it leads one to overreact, to develop and procure new military systems far in excess of what is likely in fact to be required to compensate for adversary developments. One's overreaction is likely to produce a further overreaction by one's adversary as a result of his application of worst-case analysis, and so on. Thus, in worst-case analysis one has the essential ingredient of the arms race.

The seriousness of the problem is often obscured by what York calls "the fallacy of the last move," the belief that one's adversary will in fact not react to one's own decisions. This is perhaps illustrated by the views of most of the members of our Joint Chiefs of Staff over the last few years regarding a possible deployment by the United States of a large scale nationwide ABM program. Their testimony suggests that they felt such a deployment would be worthwhile simply because it could save lives and property. They seemed to discount almost totally, particularly during the early sixties, the possibility of the Soviet Union's offsetting our defense by improving the offense. Soviet military planners presumably have discounted U.S. reactions, as they went ahead with their ABM programs.

Finally, the problem of worst-case analysis is exacerbated if lead times required for the development and procurement of strategic systems are long. In that case the uncertainties involved in arms-race decision-making will be increased. In deciding whether to develop and deploy strategic weapons, a government, to the extent to which it is attempting to be responsive to actions of other countries, is forced to ask for intelligence predictions up to ten years in advance. The range of uncertainty must increase toward the future and any decision-maker who is responsible for national security will probably make judgments based on the most unfavorable predictions within that range. By so doing the decision-maker is more likely reacting not to what the other country *will* do but to what it *might* do. The MIRV and ABM systems have precisely this characteristic of a long lead time, which compounds the other kinds of uncertainties already inherent in these systems.

With the recognition of the pernicious effects of worst-case analysis it is apparent that arms-control efforts ought to be focused strongly on preventing specifically those kinds of developments which most lend themselves to just such analysis. That is why Herb York has identified ABM systems and MIRVs (including the developing of improved accuracy and reliability for the latter) as *the* critical destabilizing developments on the horizon. That is why stopping ABM and MIRV programs have recently commanded such attention in the arms-control community, and why stopping them is so much more important than limiting, say, weapons yields or numbers of bombers.

The problem is in the fact that both ABM and MIRV systems lend themselves to a remarkable degree to worst-case analysis. In each case there is likely to be considerable uncertainty about effectiveness: in the case of the ABM because of uncertainty about adversary capabilities and tactics and because there are so many things that can go wrong with such a complicated system; and in the case of MIRVs because one may not know how many warheads the adversary has on each missile, how accurately they can be delivered, or with what reliability. Moreover, with respect to both MIRVs and the ABM systems, there are likely to be ambiguities as to the purpose of deployment. In the case of MIRVs, it will not be clear whether they are being deployed to facilitate penetration of possible adversary ABM defenses, for use as a counterforce weapon so that one can more effectively destroy fixed adversary missile forces by preemptive attack, to permit the attack of a very large number of targets, or for all three purposes. In the case of ABM systems, deployment may be to defend retaliatory capabilities, a stabilizing move; it may be to defend population against a retaliatory strike by the adversary—a prospect which would likely stimulate a buildup in adversary strategic offensive strength; or it may be for still other purposes, such as defense against lesser powers or to cope with accidentally launched missiles.

Because we lean so heavily on worst-case analysis, not only are we confronted with a situation of arms race instability, but we may possibly in the future find ourselves in a situation of increasing crisis instability as well.

Assume both superpowers deploy ABM systems of some sort and install MIRVs of high accuracy and reliability on their strategic missiles. In such a situation, and with the application of worst-case analysis, it is very likely that one side or the other would conclude that its land-based missile force would be vulnerable to preemptive attack by its adversary's MIRVs, and that its adversary's ABM and air defense systems could then cope with residual retaliatory forces. Indeed, we are seeing just such a reaction in the United States right now, even prior to the realization in the Soviet Union of MIRVs of high accuracy. While the reaction in the United States has so far focused on defending or further hardening our ICBMs, on adding MIRVs to our offensive missiles, and on developing new strategic offensive systems, another kind of reaction is also plausible; that is, that a decision would be made that it would be necessary to provide for launch of ICBMs based not on the impact of adversary missiles but rather on radar warning of an attack. Were this latter response to occur, worst-case analysis would have led to the realization of York's ultimate absurdity. The fate of both the Soviet Union and the United States would then be dependent on the performance of electronic devices—radars and computers—and on technicians and junior officers who would have the responsibility for programming the computers and for maintaining the electronics. There would be little, if any, opportunity for political leadership to bring its judgment to bear in deciding whether or not ICBMs should be launched.

Such a delegation of decision-making responsibility by political leadership to subordinates and machines could plausibly occur either during original systems design or afterwards if the worst-case analysts were persuasive in their arguments that extremely short reaction times would be required.

Then there could be a failure of the systems, particularly in a time of crisis—a launch of ICBMs based on ambiguous or false indications of adversary attack.

Is such a sequence likely? Perhaps. At least it seems more likely that thermonuclear war would be initiated in this way than if the political leadership retains in its own hands the capability and the responsibility for making assessments regarding indications of adversary attack and for launching an attack of its own.

In conclusion, let me just state that I am personally persuaded of the legitimacy of concern not only about York's first absurdity but also about his second. The latter is, however, more contentious. Surely one of our purposes here should be to explore just how serious a possibility it is.

Assuming York is right, it would seem worth looking to technology to see if it can in any way help us avoid a further erosion in political control over doomsday.

And we should obviously also be trying to identify future technical developments that may lend themselves particularly to worst-case analysis and which, therefore, may be destabilizing in one or the other of the senses that I have been discussing. With early identification, there may be some hope of avoidance or effective control.

Summary of Discussion

Instability Caused by ABM Deployment. An ABM system which was unambiguously and solely for defense of ground-based missiles would not be destabilizing. The Safeguard ABM system, however, does not have these characteristics and is probably not the best of the many options available to maintain a secure deterrent. Furthermore, an effective anti-Chinese ABM system might even be stabilizing if it could be made clear that such a system did not foreshadow an ABM system effective against the other superpower's force. However, because of the intrinsic uncertainties in performance, this is not possible. The extent of deployment necessary for a high-confidence anti-Chinese ABM system would so shorten the lead time for a massive deployment that the other superpower side would almost certainly respond by increasing its offensive capability.

Instability Caused by MIRV Deployment. If neither side had fixed land-based missiles then MIRV deployment would not be destabilizing. But at present both superpowers have most of their deterrent in the form of fixed land-based missiles and hence the deployment of MIRV by one side is likely to lead to response by the other and therefore is destabilizing in the arms-race sense.

The characteristic of MIRV deployment that makes it more destabilizing than separate missiles is the concentration of warheads at a single aim point. Since separate missiles are generally likely to be spread over a distance larger than the assumed kill radius of the attacking missile, in the case of such deployment at most one missile can be destroyed by each incoming warhead. If neither side's missiles are MIRV'd, this means that to destroy a single missile on one side requires the expenditure of at least one missile on the other. But with the attacker MIRV'd, the exchange ratio increases, so that with one attacking missile, more than one of the opponent's missiles can be destroyed. The degree to which this situation leads to instability is a function of accuracy. The more accurate the attacker's missiles and the greater the numbers of warheads per MIRV'd missile, the smaller fraction of his missiles he needs to expend in order to neutralize his opponent's force, and hence the greater his potential advantage following his counterforce attack. Except to the extent that there exist nontargetable, submarine-based or mobile land-based missiles, this situation provides an incentive for a preemptive attack.

Crisis Instability Caused by MIRV and ABM Deployment. The nature of the crisis instability that may result from the deployment of high-accuracy MIRVs and ABM defense can be described as follows: Although no decision-maker who wanted to avoid all damage to his homeland would strike first in a crisis, it is conceivable that under circumstances where it might appear that the outbreak of war is likely, damage to one's homeland may be significantly reduced by striking first. To the extent that such incentive to strike first does exist, the technological innovations would have resulted in a "crisis-unstable" balance.

Flexibility of Strategic Spending. The amount that each of the superpowers spends on strategic systems is fairly flexible and not limited by the overall economic restraints on the total defense budget. In the United States this is true because the strategic portion of the budget accounts for about one-quarter of the total defense budget, not including the AEC. In fact, at the moment, although the defense budget is declining in the United States, the amount expended for strategic forces is rising.

Self-destruct Mechanisms for Missiles. It was suggested that the danger of a launch-on-warning policy may be somewhat reduced by installing self-destruct mechanisms in the missile. Several drawbacks were suggested to such mechanisms:

1. They may lead to an erosion of the safeguards preventing undesirable launch.
2. If the missiles are launched on false warning, the destruct mechanisms may fail.
3. Even if the mechanisms did not fail, the country would then be left with no, or at least a degraded, deterrent force.

Historical Context of Worst-case Analysis. It was remarked that worst-case analysis, although presently characteristic of much military planning, would have been quite foreign to military leaders until the 1940s. For example, Elting Morrison has described the reluctance of navies during the nineteenth century to adopt new weapons and their complacency in the face of advances in naval technology made by their rivals. Possibly this shift was caused by the revolution in the art of war marked by the advent of nuclear weapons.

1 New Technology

Ballistic-missile Guidance

D. G. Hoag

World governments and their citizens are now engaged in a serious and urgent examination of courses of action to stabilize the strategic-arms race, to achieve some measure of arms limitation, and to avoid the dreadful possibility of a nuclear-weapons exchange between nations. Political, social, and technological factors complicate these discussions. One aspect of technology which plays a prominent role is that of the operational and performance capabilities of long-range ballistic-missile guidance.

Ballistic missiles are those which are thrown with sufficient velocity to travel without further propulsion to the target. The initial velocity of such a weapon can be imparted by a strong arm, a medieval catapult, or by the explosion of gunpowder in a cannon. But the ballistic missiles we examine here are modern weapons with nuclear warheads launched by booster rockets on free-coasting flights of thousands of miles.

The factors of interest of ballistic-missile guidance which will be covered herein focus on accuracy and operational characteristics. We start with descriptions of a ballistic missile, of inertial sensing, of inertial navigation, of inertial guidance,

D. G. Hoag is the Director of the Apollo Guidance and Navigation Program.
The author extends thanks to Mr. William Stameris and Mr. Robert Weatherbee for their editorial advice and assistance. Permission to reproduce Figure 8 was generously given by Mr. Milton Trageser and Mr. John Dahlen.

and of rocket thrust-vector control. The informed reader may want to skip this material and pass on to the recapitulation on page 36. General free-flight ballistic-trajectory characteristics are then described, followed by discussions relating guidance-system errors to target miss and sections describing performance-improving and operational-flexibility options for ballistic-missile guidance. Finally, concluding that there are no real technological limits to achieving significant improvement in ballistic-missile accuracy, the implications of this fact are examined.

The Ballistic Missile

In its simplest form a modern ballistic-missile weapon system has two major parts: (1) a booster rocket of one or more stages, and (2) a warhead housed in a protective structure of particular shape to survive a high-speed reentry passage back through the atmosphere.

Three distinct phases of flight are evident: (1) powered flight or rocket boost from launch, which accelerates the missile upward through the earth's atmosphere and downrange toward the target, to rocket thrust termination at the warhead release point at high velocity above the atmosphere; (2) ballistic coast of the warhead in a free-fall trajectory in the vacuum of space; and (3) reentry of the warhead back through the atmosphere.

The motion of the weapon through these three phases is determined by the forces acting on it. The payload is accelerated by the large propulsive force of the rocket in a direction opposite to that of the exhausted gases of the burning propellant. During the atmospheric passage of the rocket-boost and warhead-reentry phases, the motion is affected by large aerodynamic forces of drag and lift. And during all phases, the vehicle is pulled towards the center of the earth by the force of gravitation.

Only the rocket propulsive force is under direct control to affect the aim of the weapon. The accuracy of a ballistic missile depends upon the ability to steer the rocket thrust so that the position, speed, and direction of motion at thrust termination are precisely those needed to establish a trajectory which will hit the target. But the accuracy also depends upon the ability to predict the motions in the free-coasting trajectory and the ability to specify target coordinates accurately. We will examine these factors.

The guidance of a ballistic-missile booster rocket requires the accurate measurement of motions as they occur during rocket-powered flight. The measurements can be made either by radio or radar techniques to measure the motions directly, or by so-called inertial-sensing techniques to measure the forces causing the motion.

Radiation sensing using radar stations and radio-ranging equipment located on accurate baseline arrays interrogating cooperative transponder beacons on the missile can provide almost micrometer precision in the measurement of location and velocity of the burning rocket. Some characteristics of these radiation-sensing methods are undesirable, especially for a weapon system. An obvious drawback is that the ground stations are complex and conspicuous and are therefore vulnerable to enemy countermeasure activity. For this reason the guidance of modern ballistic-missile rockets has depended upon and caused remarkable advances in the technology of inertial sensing. Inertial sensing is completely self-contained, having no active or passive radiation contact with the outside world.

Inertial Sensing

The principle of inertial sensing depends upon the direct measurement of forces acting on a test mass inside an instrument known as an accelerometer. There are two forces acting on the accelerometer's test mass: the force of gravity and the force transmitted by the support that constrains the test mass so that it shares the motion of the accelerometer and that of the missile carrying the accelerometer. As both of these forces have direction and magnitude, they must be treated analytically as vectors.

The vector force of gravity, $\mathbf{f}_g$, on the test mass is proportional to both the magnitude of the test mass, m, and to the magnitude of the local acceleration of gravity, $\mathbf{g}$, and has the direction of the local force of gravity:

$$\mathbf{f}_g = m\mathbf{g}. \quad (1)$$

The vector force of the test-mass support, $\mathbf{f}_a$, in the accelerometer must be measured by components within the instrument. There are many accelerometer configurations. One of these is conceptually simple: the test mass is supported by calibrated beams with strain gages that measure the components of the support force and provide the accelerometer output signal.

The vector sum of the gravity and accelerometer-support forces is the total force on the test mass causing its motion and is proportional to the net acceleration of the test mass:

$$\mathbf{f}_g + \mathbf{f}_a = m\mathbf{a}. \tag{2}$$

Solving for this acceleration and substituting Equation 1 gives:

$$\mathbf{a} = \frac{\mathbf{f}_a}{m} + \mathbf{g}. \tag{3}$$

This is the fundamental equation of inertial sensing. The quantity **a** is the acceleration of the test mass and therefore is the acceleration of the missile; **g** is the local acceleration of gravity which is an accurately known function of the position of the vehicle relative to the gravitating masses of the universe. Near the earth,

$$\mathbf{g} \sim -\frac{GM_e}{r^3}\mathbf{r}, \tag{4}$$

where GM_e is the earth's gravitational constant and **r** is the position vector of the missile relative to the earth's center.

The term $\mathbf{f}_a/m$ in Equation 3 is the measured specific force (force per unit mass) in the accelerometer applied to the test mass by the support and has units of acceleration. It is easily shown that this specific force equals that for any mass in the missile under the assumption that the missile is not rotating. Indeed, this specific force measured in the accelerometer equals that for the missile as a whole, being equal to the ratio of the vector sum of all the direct applied forces to the total mass of the missile. The direct applied forces are those from the rocket propulsion, aerodynamic lift and drag, and any other miscellaneous externally applied forces except that of gravity. From a different point of view, the specific force measured in the accelerometer is equal to that component of missile acceleration arising from all the applied physical forces but not gravity. While sitting on the earth before launch the acceleration is zero, but the accelerometer will measure the specific force of the launch pad pushing up on the bottom of the missile with the magnitude equal to the acceleration of gravity. Later, after the rocket shuts down and is in free-fall coasting flight when gravity is the only force, the accelerometer will read zero even though the equipment is accelerating due to the force of gravity.

This important point of inertial guidance must be emphasized: the accelerometer cannot measure the component of acceleration due to the force of gravity. This part of the motion must be determined analytically from the known magnitude and direction of gravity as a function of position.

In explaining the nature of the specific force in the previous paragraph it was assumed that the missile was not rotating. This was done so that the centripetal forces would be zero. A rotating but rigid missile has different instantaneous accelerations at different locations. But our guidance sensors need only measure specific force at a local spot in the vehicle, being assured that the average velocity and position of other missile locations will not be much different. In fact it is advantageous to locate the accelerometers at the top of the missile near the warhead so that warhead release will be based upon its local motion rather than that of the center of gravity of the whole rocket.

From the basic equation of inertial sensing, Equation 3, the total local acceleration is determined as the vector sum of the accelerometer-measured specific force and the analytically determined acceleration of gravity, such as given in Equation 4, as a function of missile position. This equation can be integrated once to get velocity and once again for position with due regard for initial conditions of velocity and position as embodied in the constants of integration.

At this point, some of the practical aspects of implementation are introduced. Equation 3 is a vector equation. Equipment to deal with this equation will operate on vector components represented in an appropriate reference coordinate frame. Suitable reference coordinate frames which make Equation 3 true are nonrotating and nonaccelerated (in the sense that the center of the earth is nonaccelerated, being in free fall in its motion in the solar system). For ballistic-missile inertial guidance it is common practice to implement the acceleration equation in an earth-centered, nonrotating coordinate frame. It is also common practice to mount the accelerometers on a gyroscopically stabilized platform so that the accelerometer specific-force output signals are directly components of acceleration in a nonrotating coordinate frame.

Gyroscopes, by the action of conservation of angular momentum of their rapidly spinning wheels, can generate signals proportional to the angular displacement in space of their cases. If the gyros are suitably mounted on a stable platform such as the three-degree-of-freedom gimbal-supported member represented by Figure 1, the gyro signals become attitude error signals for servo electronics which can then drive gimbal-axis torque motors to keep the inner gimbal spatially nonrotating in spite of the rotations of the missile and stable-platform support structure. It will hold that orientation in space determined by an initial alignment process accomplished on the ground before launch. The angular rotations measured by mechanical-to-electrical-angle signal generators mounted on each axis of the gimbal system are the components of missile angular motion with respect to the reference orientation determined

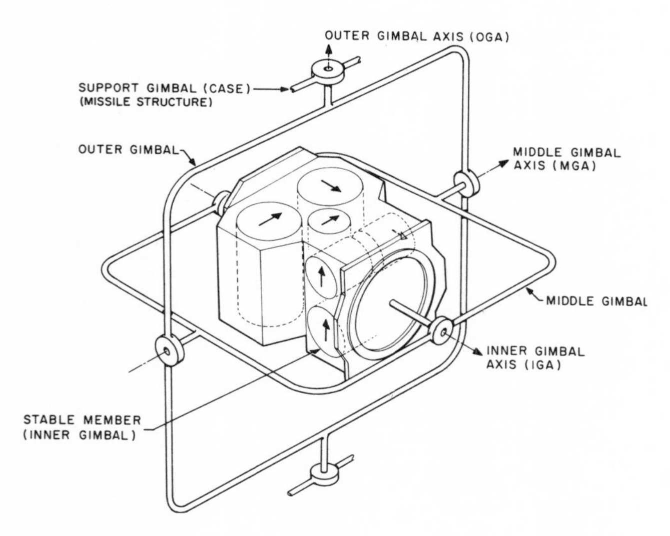

Figure 1
Simplified diagram of a stable platform.

by the prelaunch alignment. These are the rotational outputs (roll, pitch, and yaw) of the inertial sensor. The translational outputs are the components of specific force measured by the accelerometers mounted on the nonrotating platform.

Gyroscopes and accelerometers, then, are the basic instruments of inertial sensing. There have been many successful design approaches of these instruments incorporating a variety of ingenious features motivated towards optimizing various measures of performance. Figure 2 is a cutaway view of the single-axis gyroscope used in Apollo-spacecraft guidance. (The term "single-axis" refers to the fact that the gyro is sensitive to input angular

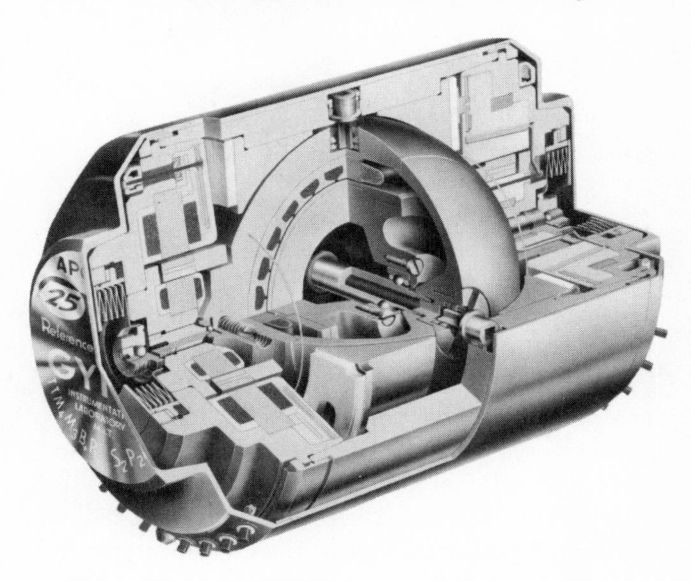

Figure 2
Single-axis gyroscope.

Figure 3
Single-axis pendulous accelerometer.

motion only about a single axis.) Three of these with input axes orthogonally oriented are required to stabilize a platform. Figure 3 shows a cutaway view of the single-axis pendulous accelerometer also used in Apollo guidance. The test mass is configured as a single-axis pendulum restrained by torque feedback to a reference angular position with respect to the case. The torque required is the measure of specific force. Three of these instruments with input axes orthogonally oriented are needed to measure all components of the specific force. The implementation of a complete gimbal-supported stable-platform inertial sensor is shown in Figure 4.

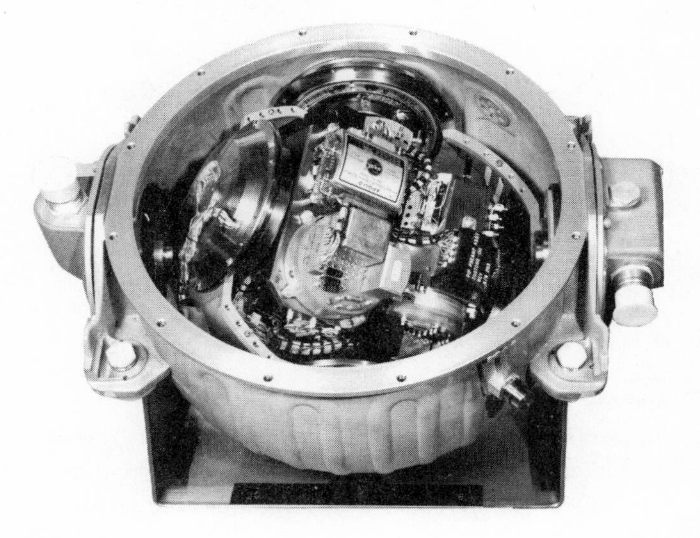

Figure 4
Stable platform.

Inertial Navigation

The use of inertial sensing to guide a missile rocket can take a number of different forms. A conceptually simple form is chosen here for illustration. Several steps are involved. The first step integrates the basic equation of inertial sensing to produce explicit signals representing missile translational velocity and position. This process is called inertial navigation and is illustrated in Figure 5. Since the accelerometers do not measure the components of acceleration due to gravity forces, these are separately computed and added to the accelerometer signals as shown. The integration of this net acceleration, the left side of Equation 3, is the velocity change which, when added to the initial velocity at launch, when the

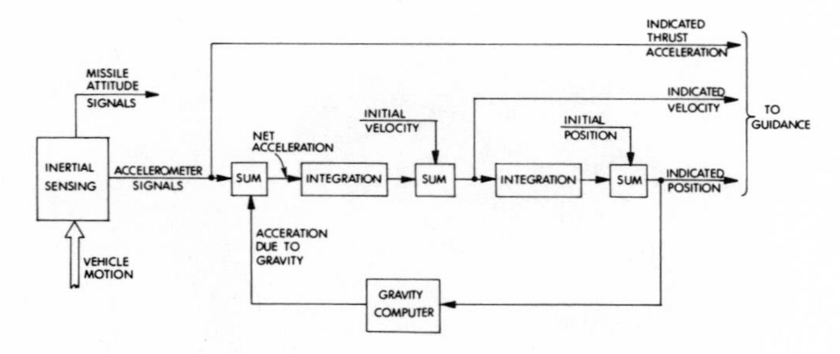

Figure 5
Inertial navigation.

process is started, is the indicated velocity output signal. The integration of the indicated velocity gives the indicated position change which when added to the initial position results in the net indicated position signal from inertial navigation. Since local gravity acceleration is a known function of missile position as given in Equation 4, then the indicated position is the necessary input to the gravity acceleration computer, forming an interesting feedback as shown. All variables are vectors, but actual implementation, of course, is performed by means of the vector components.

There are three outputs of this inertial navigation: (1) inertially indicated specific force used as a signal proportional to rocket thrust acceleration, (2) indicated velocity, and (3) indicated position. All three are used in the inertial guidance function described next.

Inertial Guidance

The function of inertial guidance is to steer the rocket thrust to achieve the desired terminal condition.

During passage of a ballistic-missile rocket up through the atmosphere the objective of steering is to control the motion so as to pass safely through the period of high aerodynamic loading. During this period the missile can tolerate only a small angle of attack and still maintain control and structural integrity. It is usual, then, to steer this early phase through an angle history which is a predetermined function of time from launch. Once safely out of the atmosphere the guidance steering is then motivated to reach computed rocket-cutoff conditions.

Figure 6 represents a block diagram for one concept of ballistic-missile inertial guidance. Note that the previous figure for inertial navigation is abbreviated and represented by the box on the left of Figure 6. Guidance must steer the rocket so that

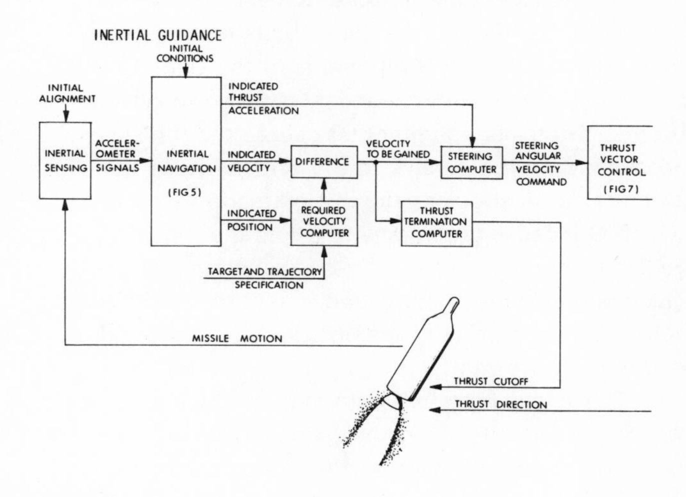

Figure 6
Inertial guidance.

the warhead will be released on one of the many trajectories which will hit the target. The target is specified to the system by the input to the box labeled "required velocity computer." The function of this box is to compute the required velocity vector the missile should have at the present position and instant of time so that the warhead, if released to coast freely, would be on a trajectory having the desired properties. The difference between the required velocity and the present indicated velocity is the velocity to be gained by the rocket. If the steering of the rocket is proper, this velocity to be gained will be reduced toward zero. When zero is achieved the rocket engine thrust should be cut off as shown by sending the thrust termination signal. The steering strategy illustrated aims the missile so that the direction of the present measured thrust acceleration is parallel to the direction of the velocity to be gained, thus assuring that all components of the velocity to be gained will be simultaneously reduced to zero as desired. The steering computer causes the thrust acceleration to be parallel to the velocity to be gained by creating an angular velocity command vector which will cause the missile to rotate the thrust vector toward the velocity-to-be-gained vector. This angular velocity command is sent to the missile's autopilot or thrust-vector control system described in the next section.

This is one representation of a particular guidance scheme called velocity-to-be-gained cross-product steering. The term "cross-product" arises from the use of a vector-cross-product implementation in the steering computer to derive an angular velocity command signal proportional to the angle between the indicated thrust acceleration and the velocity-to-be-gained vectors.

This is an appropriate place to comment on the powerful implications of the simple event of the guidance system thrust termination signal. A ballistic missile with its high-performance rocket stages and associated equipment depends upon many complex systems working successfully in order to complete its mission. If the guidance system senses that all three components of the velocity to be gained have simultaneously passed sufficiently close to zero so as to generate the thrust termination signal, then almost no failure can have occurred anywhere in the whole missile system up to that time that will affect the performance of the weapon significantly. The only outright failure that would mislead such interpretation is the logic and circuitry generating the thrust termination signal itself. Once the signal is generated, one is assured that everything else worked as intended including most of the guidance system as well. The event of three components of velocity arriving at zero simultaneously with a significant failure in either achieving the velocity components correctly or in measuring them improperly is near impossible. Large degradation in performance, not outright failure, in the inertial sensing instruments could fool this criterion of indicated success, but experience with these instruments and simple analysis show that such a situation is most rare compared to the many other possible failure modes elsewhere in the missile system. If the thrust termination signal is generated, and if the thrust actually ceases, as can be sensed by the accelerometers, then the weapon has been sent successfully on its intended free-fall ballistic trajectory. Such a signal can be used to arm the warhead with the assurance that there is negligible chance that it will go awry and explode far from the intended target. If the signal is not generated, a simple one-bit coded signal could be sent to the ground to indicate failure and the need for another launch against that target.

Thrust-vector Control

Thrust-vector control is the closed loop process which (1) keeps the vehicle from tumbling under the high forces of engine thrust and (2) accepts turning or guidance steering commands to change the direction of the applied rocket acceleration. For liquid-fueled rockets the engine is relatively small and control torques are achieved by the use of an engine swivel or gimbal arrangement to deflect these forces from the center of mass. For solid-fueled engines the thrust chamber contains all the unburned propellant, hence a gimbal mounting of the whole chamber is not practical. Other means of torque control such as jet vanes, jetavators, nozzle swivel, or gas injection can deflect the hot gas stream without moving the thrust chamber itself.

These methods provide control torques about pitch and yaw axes perpendicular to the long axis (roll axis) of the rocket. Multiple engines or nozzles at the base of the rocket are common and the thrust from each can be offset differentially to provide torque control about the roll axis. Otherwise special

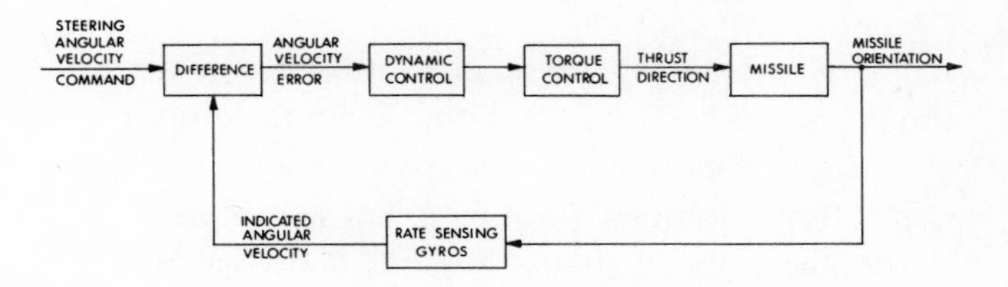

Figure 7
Thrust-vector control.

small thrusters can be arrayed on the side of the missile for roll control.

A typical thrust-vector control loop is illustrated in Figure 7. The input to this system is the signal from guidance proportional to the desired angular velocity of the thrust to rotate the missile so as to reduce the thrust-direction error toward zero. This angular velocity command is compared with the actual missile angular velocity presently existing as measured by rate gyroscopes. The difference becomes the missile angular-velocity error, which is amplified and treated with dynamic compensation filtering in the box labeled dynamic control. The output of this box is the command to the torque control deflecting the thrust such as described above. The missile responds to the applied torque according to its moments of inertia and other dynamic properties. The resulting angular motion is sensed by the rate gyroscopes to close the loop. Alternately, these rate gyros can be eliminated by appropriately using the missile-attitude signals from the inertial-sensing platform described earlier.

The dynamic compensation of such loops can be quite a complex design process. The missile can rarely be treated as a rigid body in the design and the effects of body bending and fuel and oxidizer sloshing in their tanks can cause large and usually destabilizing torques. Another effect with a very descriptive name is that of engine "tail wagging" where large, gimballed engines have themselves moments of inertia comparable to the rest of the missile. Commands to move the engine (the tail) result in considerable motion of the missile itself (the dog). And finally, in the flight through the atmosphere the effects of aerodynamic forces are considerable and destabilizing.

The thrust-vector-control design must recognize these destabilizing effects so that the dynamic error in following guidance commands is sufficiently small. This requirement is particularly tight near thrust termination. If the missile has high angular velocities at this time or is not following steering commands accurately, then all three components of the desired velocity will not simultaneously pass through their correct values for engine cutoff. Residual velocity error causes miss at the target if not subsequently corrected.

Recapitulation

A ballistic missile is propelled by a booster rocket to high velocity above the atmosphere where the warhead is released to coast freely in a ballistic path before it reenters the atmosphere. Guidance of this weapon to hit a specific target occurs only during the rocket boost phase which extends over only the first few minutes of flight. Guidance measurements utilize inertial-sensing techniques. Inertial sensing is the application of Newton's laws as they operate in gyroscopes and accelerometers to measure all components of missile rotational motion changes and all components of missile translational motion changes due to all forces on the missile except gravity. Navigating or determining missile position and velocity by inertial sensing requires the analytical determination of that component of missile motion due to the force of gravity acting upon it. Moreover, the inertial sensing can measure only changes in the missile motion, and the initial conditions of position (launch location) and velocity must be provided externally before launch. Guidance calculations determine the direction toward which to steer the rocket so that the velocity and position approach those conditions compatible with a free-fall coast to the target identified to the guidance system at launch. The rocket autopilot thrust-vector-control system responds to the guidance steering commands to control the direction of rocket thrust with respect to the missile in a stable fashion. When the proper conditions of position and velocity are achieved, the guidance system signals termination of rocket propulsion and the warhead is released. Now flying under the influence of gravity alone, the warhead coasts in the vacuum of space in an elliptical path until it crashes into the earth's atmosphere where it will be slowed considerably before being triggered to explode at the target.

Free-flight Ballistic Trajectory Characteristics

The free coasting phase of the trajectory approximates a Keplerian ellipse. This assumes a uniform central-force field of the earth's gravity. The plane of motion contains the present position and the center of the earth, and if the trajectory is satisfactory will also contain the location of the target at the time of arrival of the warhead. From a given warhead release point there exists a first-order infinity of free-coasting trajectories having various launch angles with respect to the horizontal and various maximum altitudes which will intercept a given target.

From this infinite set of possible trajectories, one, called the minimum-energy trajectory, is unique in that it requires the smallest velocity possible to hit the target from a given warhead release point. For a given booster rocket and payload weight, the maximum range capable of the weapon occurs with a minimum energy trajectory which corresponds to the burnout velocity of the rocket exhausted of its fuel. With this system the weapon designer has a choice in selecting trajectories for shorter-range targets. Low trajectories will have a shorter time of flight and will arrive at the target at a shallow angle which might provide an advantage in surprising the defense. With trajectories somewhat higher than the minimum energy condition, the aerodynamic heating of the reentering body is less and the miss due to non-nominal entry conditions is reduced. Moreover, the target miss of these lofted trajectories is less sensitive to errors in the magnitude of the warhead release velocity. But the miss due to errors in the direction of the warhead release velocity measured in the vertical plane are smallest for the minimum energy trajectory. And the shallow trajectories are better in minimizing the effect of time-of-flight variations on target miss.

Clearly, no general rule identifying an optimum choice can be formulated. But for ranges shorter than the maximum capability of the rocket, the weapon designer can utilize the variations in time of flight and approach angle to achieve some operational advantage. For instance, a salvo fire or a deliberate coordinated launch plan can use this freedom in choice of trajectory to cause the warheads to arrive in the same target area at carefully spaced times and at different angles so as to confuse, saturate, or exhaust the defense. The flexibility possible is illustrated in Table 1 for a constant target range of 6,700 kilometers.

The relationships among the variables of interest of all possible free-coasting ballistic-missile trajectories are summarized in Figure 8. These curves are

Table 1 Variations in Ballistic Trajectories for Constant Range.

	High Trajectory	Minimum Energy	Low Trajectory
Target Range (kilometers)	6,700	6,700	6,700
Ballistic Velocity (meters/sec)	7,000	6,400	7,000
Entry Angle from Horizon (degrees)	50	30	10
Time of Flight (minutes)	42	25	17

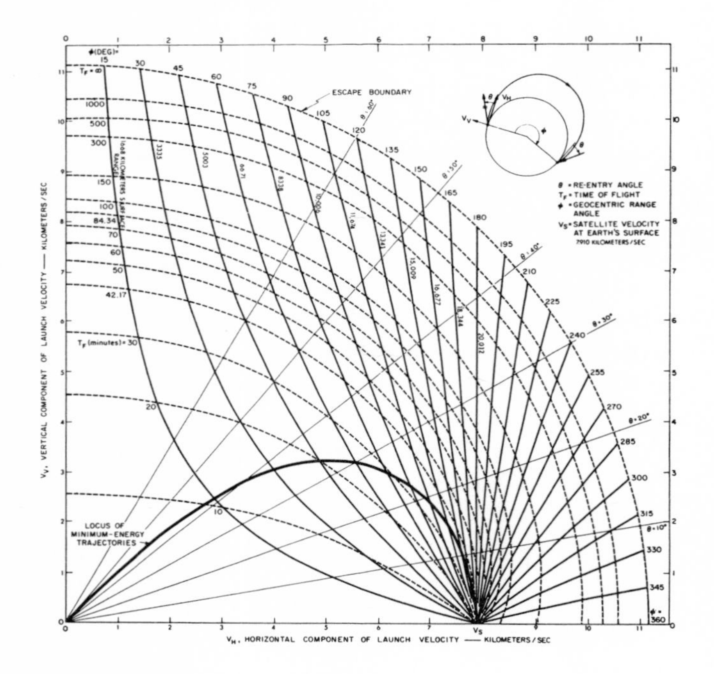

Figure 8
Free-coasting ballistic trajectories.

correct for a nonrotating spherical earth with no atmosphere and with the warhead release point at the earth's surface. In spite of these qualifications the relationships are close to those for the real case. The figure has the vertical component of warhead release velocity as an ordinate and the horizontal component of release velocity as an abscissa. Each point in the plane corresponds to a specific trajectory. Contours of constant time of flight and constant target geocentric range angle are given.

An intermediate range ballistic missile (IRBM)

might correspond to a surface range of 3,300 kilometers, a geocentric range angle of 30 degrees, a minimum-energy warhead release velocity of 5,000 meters/sec at an angle of about 37 degrees above the horizon. Flight time is about 15 minutes.

An intercontinental ballistic missile (ICBM) is generally considered to have a surface range of the order of 10,000 kilometers corresponding to a geocentric range angle of 90 degrees. At minimum energy the warhead release velocity is about 7,200 meters/sec at an angle of about 22 degrees above the horizon. Flight time is about 30 minutes.

Longer ranges are possible. A 20,000-kilometer surface range reaches exactly halfway around the earth and is the trivial point where the minimum-energy velocity angle reduces to zero. The required velocity is the earth's surface satellite velocity of about 8,000 meters/sec. This range at minimum energy is impossible if only for the presence of the earth's atmosphere, and a lofted trajectory is required with only a very slightly larger velocity needed. A weapon of this capability could reach any target on the earth.

Even longer ranges can be considered, covering more than halfway around the earth. In such a case the concept of minimum energy has no practical significance and all would be lofted to keep the angle at reentry sufficiently steep. Such trajectories might provide some tactical advantage in that the warhead would complicate the defense by approaching the target opposite from the expected direction. A target range of 10,000 kilometers (90 degrees) requires 7,000 meters/sec the short way. The long way around to the same target would be 30,000 kilometers (270 degrees) and would need a velocity somewhere over 9,000 meters/sec. This extra velocity would require a more protective heat shield for reentry and would require a much larger rocket or the same rocket would require reducing the warhead payload weight by several times. A given accuracy of guidance equipment could result in target miss not much larger, perhaps twice, on the long-way-around trip than would appear on the direct flight.

Target Miss and the Effect of Warhead-release Errors

In sections to follow we will examine many sources of error which cause miss at the target. In studying the performance of a precision guidance system, the approximations from linearizing small perturbations from a given trajectory are quite accurate. The size of a given miss effect is proportional to the size of the source error. Twice the source error means twice the miss effect. These effects are then conveniently expressed as error partials: the ratio of effect per unit of error source. Also, if several error sources affect the same variable, then the net effect of a collection of error sources is the vector sum of the individual effects of each source one at a time. If the error terms are chosen to represent the standard deviation, or root mean square (RMS), of the statistical distribution of that parameter in an ensemble of similar systems, then the net effect of a collection of error sources is the square root of the sum of the squares (RSS) of the individual contributors. This assumes that the error sources are statistically uncorrelated. Even though the statistical distribution of the individual sources may be far from normal, the resulting RSS combination of a number of contributors tends toward a gaussian distribution. The separate range and cross-range miss terms may be combined into a circular error probability (CEP), this being the radius of the circle centered on the target containing half the population of range and track miss events. The CEP is approximately equal to 59% of the sum of the range and cross-range miss terms. When the ratio of the two is far from unity this approximation fails and other equations must be used.

A large class of the target miss error sources act first to cause corresponding errors in the position and velocity at the point of warhead release. The propagation of these errors at warhead release to the miss components at the target is of interest. The desired relationships are a strong function of the target range and the particular trajectory intended. For the sake of simplicity we will use only the 10,000-kilometer range minimum-energy ballistic trajectory in what follows. Unfortunately the extrapolation to other trajectories and target ranges is not simple. It is generally true that shorter-range targets are less sensitive to error sources than are longer ranges, but there are exceptions for some error components.

Table 2 lists the target miss components for a 10,000-kilometer range, minimum-energy trajectory per unit component of the velocity and position errors at the warhead-release point. Velocity and position errors at warhead release are in vertical and horizontal components in and perpendicular to the trajectory plane. The miss is described by local

Table 2 Warhead-release and Target-miss Error Partials (10,000-Kilometer Minimum-energy Trajectory).

Errors at Warhead Release		Horizontal Miss at Target: Range (in trajectory plane)	Lateral or Track (out of plane)	Westward (due to earth rotation)
Velocity Errors at Warhead Release	Vertical	$2{,}000 \frac{\text{meters}}{\text{meter/sec}}$		$300 \cos(\text{latitude}) \frac{\text{meters}}{\text{meter/sec}}$
	Horizontal (in plane)	$6{,}000 \frac{\text{meters}}{\text{meter/sec}}$		$450 \cos(\text{latitude}) \frac{\text{meters}}{\text{meter/sec}}$
	Horizontal (out of plane)		$1{,}000 \frac{\text{meters}}{\text{meter/sec}}$	
Position Errors at Warhead Release	Vertical	$3.5 \frac{\text{meters}}{\text{meter}}$		negligible
	Horizontal (in plane)	$1 \frac{\text{meter}}{\text{meter}}$		
	Horizontal (out of plane)		(cosine of geocentric range angle $= \cos 90^\circ = 0) \frac{\text{meter}}{\text{meter}}$	

horizontal components at the target as appropriate: in the plane of the trajectory, perpendicular to the plane of the trajectory, or in an east-west direction. The vertical component of target miss is included in its effect on horizontal miss. The performance of the fusing system will determine the local altitude of the explosion.

With the generally accepted coordinate conventions all entries in Table 2 are positive. We will consistently treat them as statistical standard deviations. Moreover, all entries are rounded-off approximations. They were derived for a spherical earth with uniform central gravity field and no atmosphere and for an impulsive launch from the surface.

The right-hand column of Table 2 which lists the westward component of the miss accounts for the time-of-flight variations resulting from warhead release errors. In other words, a vertical velocity error of 1 meter/sec upward at warhead release has two effects. First, the shape of the trajectory is modified and it intersects the earth 2,000 meters further downrange and second, the time to impact the surface is increased by 0.65 second. During this 0.65 second the target, if on the equator, has moved an additional 300 meters eastward from its expected position causing this magnitude of miss to the west. If the target is not on the equator, this component of the miss is smaller by a factor of the cosine of the latitude. Since the largest miss attributable to the time-of-flight variation component is small relative to the direct-miss effect, the time-of-flight effect will be ignored in the approximations examined in later sections.

Table 2 can now be used to evaluate the miss-producing effect of errors in the guidance system as it attempts to guide the booster rocket to satisfactory warhead-release-point velocity and position components. But since the guidance performance is dependent upon the trajectory of the boost phase, we will discuss this first.

Boost-phase Trajectory Characteristics

The booster rocket of a ballistic missile may consist of two or more stages. Two stages are common. Both liquid and solid propellants have individual advantages. The ease of storing and handling in the military environment and the instant readiness for launch are strong points for the solid propellants.

Efficient rocket operation is at near-constant thrust. Therefore, as the propellant is consumed the reduction in mass causes a rise in acceleration. Peak accelerations near 10 *g* (98 meters/sec^2) are common near the burnout of each stage, but average accelerations may be only a few *g*. Thus to reach ICBM velocities near 7,000 meters/sec, the burning times are of the order of 300 seconds or 5 minutes and the rocket travels perhaps 1,000 kilometers in its curved trajectory with thrust termination occurring at several hundred kilometers altitude. Patterns significantly different than this are either inefficient or have an excessively high acceleration loading.

At launch the rocket will rise vertically for a few seconds, during which time it may rotate about its roll axis (the axis of symmetry) to put the thrust control axes in a desired orientation with the intended trajectory. After this short vertical rise and before velocity and aerodynamic loads have built up to significant levels, the rocket is given a slight pitch downrange from vertical, which starts the critical atmospheric flight phase where aerodynamic loads reach considerable magnitudes. The angle of attack must be kept small to minimize drag and heating and to assure that required control torques are within capabilities. The trajectory through the atmosphere approximates a so-called gravity turn in which the specific force of the rocket, the thrust, is aligned near parallel to the direction of motion while the force of gravity tips this direction of motion over towards the horizontal. This critical aerodynamic phase will be over before the first stage burns out above the atmosphere. Active guided steering often is not initiated until second stage flight, but the navigation of an inertially guided system must have kept track of first stage flight position and velocity changes. The guided phase in the high-altitude vacuum will allow wide differences between thrust direction and direction of motion. Steering of the thrust will be to achieve a combination of position and velocity components necessary to hit the target.

For reasons implicit in the above discussion, all powered-flight trajectories designed to release a ballistic weapon on a given free-coasting trajectory are sufficiently identical for our purposes. Any differences among them will have only small effect on our ability to relate powered-flight guidance-system source errors to errors in warhead release position and velocity.

Powered-flight Inertial-sensing Errors and Relation to Target Miss

A thorough description of the imperfect behavior of gyros and accelerometers of inertial sensing and its effect on inertial guidance can become quite complex. The gyros and accelerometers can have any of a wide variety of constructions with numerous error phenomena. The effects on inertial guidance are strong functions of the particular configuration or arrangement of the sensors with respect to the trajectory and of the physical environment of the application.

The usual procedures of error analysis involve physical testing and detailed analytical modeling so as to extrapolate the experimental data from test environments to the expected performance in the ballistic-missile powered-flight environment.

Rather than illustrate a complete analysis which would require us to consider a specific configuration and construction, we will identify the prominent effects in a more general fashion. First we will characterize the various important forms of imperfect behavior and illustrate how each might originate within the instrument to form an error component. Then, for each error component we will determine the limiting effect on the performance of the powered-flight guidance, assuming some optimization in the positioning of the inertial sensors on the stable platform. To relate error sources to thrust-termination errors we must know reasonably well the desired trajectory to be flown. Fortunately, for our purpose, as we have already argued, the powered-flight trajectory is sufficiently predictable for boost to a specific free-coasting trajectory. The resulting errors in position and velocity at powered-flight termination can be propagated to the extent of target miss using the error partials of Table 2. We examine, of course, the 10,000-kilometer range minimum-energy ballistic trajectory for which Table 2 is applicable.

The important target miss partials of each of the inertial sensing error sources as they are described below appear in Table 3.

Accelerometer Bias Error. The inertial sensing outputs for guidance are the components of acceleration

Table 3 Powered-flight Inertial-sensing Effects on Target Miss (10,000-kilometer Minimum-energy Trajectory).

Error Source (see text)	Standard Deviation Miss (meters)	
	Range	Track
Accelerometer Bias, 0.01 cm/sec^2	200	30
Accelerometer Scale Factor, 10^{-5}	600	
Accelerometer Nonorthogonality, 10 microradian	200	
Platform Servo Dynamic Misalignment 5 microradian	70	30
Gyro Bias Drift, 0.02 deg/hr	400	130
Gyro Acceleration Sensitive Drift, 0.02 deg/hr per g	700	220
Vibration Induced Gyro Compliance Drift	200	70
RSS Net Effect	1,050	270
Equivalent CEP	780 meters	

measured by the accelerometers. The first error we consider is a simple offset or bias in the individual acceleration-component indications from an ideal indication of the true component of acceleration. This can arise, for instance, from a fixed force which is not part of the force transmitted by the test-mass support system acting on the test mass. Such a force will cause an error component in the instrument's indicated acceleration equal to the magnitude of the force divided by the size of the test mass. As a typical hypothetical example, an error force of 0.1 dyne acting on a test mass of 10 grams will cause an error of 0.01 cm/sec^2 in indicated acceleration. This is approximately 10^{-5} g where g, the earth's surface acceleration of gravity, is about 980 cm/sec^2. Such a fixed bias error in indicated acceleration will propagate to an error at thrust termination during the time of active guided flight. For a 300-second guided flight of our 10,000-km-range trajectory, this is a component of velocity error at warhead release of

$$0.01 \text{ cm/sec}^2 \times 300 \text{ sec} = 3 \text{ cm/sec}$$

in the direction of the sensitive axis of the accelerometer being considered. The corresponding error in position component at warhead release is

$$\tfrac{1}{2} \times (0.01 \text{ cm/sec}^2) \times (300 \text{ sec})^2 = 500 \text{ cm}.$$

Applying these errors to the error partials of Table 2 we see the worst effect of the velocity error would be a 180-meter miss at the target and the worst effect of the position error is a miss of less than 20 meters. For the class of errors in this section, inertial-sensing errors, the effect of corresponding position errors at thrust termination is always small with respect to the effect of the correlated velocity error. For this reason we will consider only the velocity effect in the remainder of this section.

The accelerometer bias error of 0.01 cm/sec^2 caused 3 cm/sec velocity error at thrust termination in the direction of the sensitive axis of the accelerometer under consideration. With the necessary three accelerometer-measured components each having this bias, we will have three components of velocity error at thrust termination. Assuming these are standard deviations and statistically uncorrelated, we use Table 2 and conclude that a bias error of 0.01 cm/sec^2 in all components of acceleration measurement will result in a standard deviation of target miss of about 200 meters in range and 30 meters in track. These are entered in Table 3.

Accelerometer Scale-factor Error. An error in the accelerometer indication proportional to the indication itself is called scale-factor error. It can arise, for instance, by improper calibration in the instrument's measurement of test-mass support force. The resultant error in velocity will be proportional to the velocity change sensed by the instrument. For an ICBM powered flight the cutoff velocity is about 7,000 meters/sec but the booster had been fighting unsensed gravity and the integrated output of the accelerometers is nearer 9,000 meters/sec. A scale-factor error of 1 part in 10^5 or equivalently 10 parts per million (10 ppm) would result in a thrust-termination velocity error of 0.09 meter/sec. We assumed the triad of accelerometers to have been arranged so that the sensitive axis of one picks up the major part of the average specific force. This practice would be motivated by the fact that only one instrument would require precision in the scale-factor calibration. In such a case our 10 ppm scale-factor error and 0.09 meter/sec velocity error results in a target-range miss of about 600 meters. There is no track miss.

Accelerometer Nonlinearity. A number of effects can cause the output signal of the accelerometer to be nonlinearly related to the acceleration component existing along the input axis. With modern precision inertial-grade accelerometers this effect can be kept small with respect to other errors. Such an effect, then, is not listed in Table 3.

Accelerometer Sensitive-axis Misalignment. Directional alignments of accelerometer axes of sensitivity are extremely important since a misaligned instrument will measure unexpected components of acceleration existing at right angles to the presumed direction of sensitivity. The error in direction of the sensitive axis from that direction desired or expected can arise from four sources: (1) the initial spatial alignment of the package, (2) the rotation drift due to imperfect stabilization by the gyroscopes, (3) the nonorthogonality in the mounting of the accelerometer sensitive axes one to another, and (4) the platform servomechanism angular bias during accelerated flight.

Number (1) is treated in a future section and number (2) is considered later with the various forms of gyro drift. Number (3) is that part of the alignment error which should be removed during manufacture of the stable platform. It concerns the physical adjustment of each accelerometer sensitive axis to be orthogonal to the other two. With the typical prelaunch alignment as will be described, the only significant nonorthogonality term is that of the two accelerometers in the plane of the trajectory. Assuming this nonorthogonality to be 10 microradians (2 seconds of arc), then the pertinent accelerometer will sense in error approximately 1 part in 10^5 of the integrated thrust acceleration of 9,000 meters per second. This is about 0.1-meter/sec error at thrust termination, but in a direction rather insensitive to target miss. Assuming a miss error partial of 2,000 meters per meter/sec, the resulting range miss is estimated as 200 meters and is so entered in Table 3.

Misalignment effect number (4) above, platform servo angular bias during accelerated flight, should be small with respect to prelaunch alignment, number (1). This assumes reasonably good servomechanism design, electronic noise control, and mechanical balance of the platform. Assuming a 5-microradian error (1 second of arc) in alignment about each axis, then the resulting components of miss at the target are 70 meters in range and 35 meters in track.

Gyroscope Bias Drift Rate Error. An important contributor to target miss is gyro drift, which rotates the stable platform and the sensitive axes of the accelerometers to pick up components in error of acceleration occurring normal to the initially aligned sensitive-axis directions. Like the accelerometer, gyroscopes develop error due to extraneous forces acting on the wheel support gimbal. A fixed extraneous torque on the spinning wheel gimbal support will precess the wheel, causing a fixed drift (bias drift) of the desired spatial reference at a rate equal to the torque divided by the angular momentum of the wheel. If we assume a typical hypothetical gyro with a wheel angular momentum of 10^6 dyne-cm per rad/sec then an extraneous torque of 0.1 dyne-cm will cause a drift rate of 0.1 microradian/sec. This is 0.02 deg/hr or about 1.3 meru, where a meru is a milli-earth-rate-unit, one thousandth of the angular rate of earth's rotation.

A triad of gyros with orthogonal input axes that stabilize a platform and accelerometers from rotation are here assumed each to have an uncorrelated bias drift rate standard deviation of 0.02 degree per hour. How this affects the output of the accelerometers in error depends upon the time history of the acceleration components being sensed. With our expected powered-flight trajectory and evaluation of the proper integrals we get thrust termination velocity standard deviation errors of typically 0.1 meter/sec vertical, 0.06 meter/sec horizontal in plane, and 0.13 meter/sec horizontal out of plane. We have assumed here that the gyro sensitive-input axes are parallel/perpendicular to the launch vertical and the trajectory plane. Applying these results to Table 2 we obtain about a 400-meter range miss and a 130-meter track miss at the target, which are entered in Table 3.

Gyroscope Acceleration Sensitive Drift Error. An important source of extraneous torque on the gyro wheel support gimbal is the mass-unbalance effect when the gyro is accelerated. If the center of mass is not exactly on the torque-summing output axis, an extraneous torque will develop causing gyro precession and drift. The drift rate is proportional to the degree of pendulosity and the nongravitational acceleration of the gyro case, and is inversely proportional to the wheel angular momentum. A drift-rate coefficient is defined as the drift rate per unit of acceleration. It is common practice to use *g* as the unit of acceleration in this case. By orienting the gyros such that their torque-summing output axes are parallel to the average thrust direction, the acceleration of thrust cannot make large torques from the imbalance and the effect of this error source can be minimized. However, this can be done with only two of the three instruments since at least one must have its rate-sensitive and unbalance-sensitive axis along this direction. We assume here that the instruments are parallel/perpendicular to the launch vertical and the trajectory plane and that the units have acceleration-sensitive drift-rate terms of 0.02 deg/hr per *g* standard deviation for acceleration along both the gyro wheel spin axis and the input axis of sensitivity of each gyro. Again, with our expected powered-flight trajectory and evaluation of the proper integrals, we get thrust termination velocity standard deviation errors of approximately 0.2 meter/sec vertical, 0.1 meter/sec horizontal in plane, and 0.22 meter/sec horizontal out of plane. Using the miss partials of Table 2 we obtain standard-deviation miss components of 700 meters in range and 220 meters in track, which are entered in Table 3.

Gyroscope Compliance Drift Rate Error. If the center of gravity of the supported mass is carefully adjusted to lie on the torque-summing axis, then there will be no first-order effect due to acceleration, even though the structure may deflect under the acceleration load and move the center of mass with respect to the torque-summing axis. This is because the movement is generally parallel to the direction of acceleration, and no lever arm for torque appears. However, due to mechanical dissymmetry of the wheel support structure and anisoelastic properties of the wheel bearings, the movement of the center of mass due to compliance will not necessarily be parallel to the acceleration. An effective lever arm for the displaced center of mass can thus develop. This produces an extraneous torque and gyro drift which are proportional to the square of the acceleration, since the displacement and the accelerating force are each proportional to acceleration. Any structural dissymmetry which causes this effect is a consequence of the special shapes needed to support the wheel axis normal to the torque-summing axis. The effect is usually maximum for accelerations directed 45 degrees from the spin and input axes and is null for accelerations along these axes. For many configurations the gyros will be mounted so that thrust has no components to produce compliance drift. A more important source of acceleration to excite these effects is vibration resulting from rough burning of the rocket. Since the drift is proportional to the square of acceleration, the drift does not change sign as the instantaneous direction of vibration acceleration changes sign. This rectification can produce a significant drift similar to a bias drift if the character of the vibration stays uniform. It will not stay uniform throughout the flight, and the equivalent bias drift will change magnitude and can change sign. To account for this effect we enter half the bias drift effects calculated previously into Table 3 to account for vibration-excited compliance drift.

The relationships and assumptions discussed above on the powered-flight inertial-sensing errors are summarized in Table 3. The list is incomplete; many more error sources exist in a typical inertial-sensing system. But probably the significant contributors to imperfect performance for any particular system have been included. The sizes of the error sources have not been chosen completely arbitrarily. The intent was to create a set which would result in a CEP near one kilometer and thereby illustrate the high degree of precision in sensing required to reach this level of weapon performance.

Inertial-guidance Initial Conditions and Relation to Target Miss

Before launch the guidance system needs both initialization information and targeting information. Targeting problems will be covered in a later section. The initialization is the position, velocity, and orientation information needed by the inertial-sensing and inertial-navigation apparatus.

The inertial-navigation concept shown schematically in Figure 5 showed explicitly required inputs of initial velocity and initial position which are added to the integrated inertially sensed acceleration to develop navigation parameters. For fixed-base launching the initial position is simply the location and altitude of the missile on the launcher expressed in the appropriate coordinate frame. Sufficient accuracy is not difficult to obtain. The position errors at launch propagate without significant modification to the point of warhead release. The coordinate rotation through the geocentric angle from launch to thrust termination must be recognized before using Table 2. This is perhaps 10 degrees. Note in Table 2 that an error in altitude causes 3.5 times its value in range miss. The horizontal in-plane component of error causes an equal range miss. For the 90-degree central angle chosen in our reference ballistic trajectory, position error component out of the plane of the trajectory has no first-order effect at the target. Assuming a 100-meter variance in initial-position error in each component, then the resulting range miss for our target is about 360 meters and there is no track miss.

The initial velocity for a fixed-base launch is essentially zero in an earth-fixed coordinate frame. But assuming an inertially nonrotating computing frame, the eastward velocity of the launcher due to the earth's rotation must be considered. This is 460 meters/sec for launch at the equator and is reduced by a factor of the cosine of the launch latitude for other locations. One can assume trivial errors in fixed-base launcher initial velocity. The moving-base situation discussed later is not so easy. For purposes of illustration we assume 0.1 meter/sec initial-condition velocity-error variance independently in each axis. This gives a 650-meter range miss and a 100-meter track miss.

The gyrostabilized platform carrying the accelerometers must be given an initial orientation for launch. Orientation initialization (or equivalently, platform alignment) is unlike position and velocity initialization in that it necessitates achieving a physical state of the platform. Position and velocity can be represented by numbers and the initial values can be transmitted by electrical signals to the inertial navigation computer. Although angles can be transmitted by signals, orientation in the sense required cannot be so transmitted unless there is some physical reference available from which to measure the angles when received. Some physical relation to earth references is required. Since alignments to a few seconds-of-arc accuracy are needed, the physical orientation of the missile itself as a reference as it sits in the launcher is not stable enough. Reference directions can be surveyed and then transmitted, for instance, with beams of light through a window in the missile to the base of the stable platform or to the platform itself from which the desired initial alignment can be derived from electrically transmitted angles.

The earth's gravitation vector and the earth's rotation about its axis can be used by the inertial sensing gyros and accelerometers to provide a self-alignment capability as is done with a gyrocompass. The technique applies precessional torques to the platform-stabilization gyros to orient the platform so that the sensed acceleration along two axes is nulled. When this is achieved the two accelerometer axes are horizontal and the platform has a known orientation with respect to the local vertical. Since the vertical rotates in space as the earth rotates, it is necessary to keep gyro precessional torques applied to keep the platform orientation vertical. The horizontal axis not requiring precessional rotation must be in the east-west direction. The azimuth component of the alignment reference is now achieved. This gyrocompass action works well except near the earth's poles where the vertical and earth's rotation axes are parallel and no inertially sensed azimuth is possible.

The accuracy of inertially sensed initial alignment depends upon the errors of the gyros and accelerometers. With an accelerometer bias such as assumed for powered-flight inertial sensing, 0.010 cm/sec^2, the indicated vertical will be tipped in error by an angle equal to the bias divided by the acceleration of gravity, 980 cm/sec^2. This is an initial alignment angle error from the plumb-bob vertical of 10 microradians (2 seconds of arc) about both axes. In addition, the plumb-bob vertical itself may not be parallel to the reference coordinate system vertical due to local gravity anomolies such as those described in a later section. For fixed-base launches, this deflection of the vertical can be determined and a correction applied quite accurately. If we assume another 10 microradians due to imperfect compensation for gravity deflection of the vertical, we now have a total error of 20 microradians for each axis, which with our standard 10,000-km trajectory causes a 260-meter range miss and a 90-meter track miss at the target.

The error in gyrocompass azimuth alignment will depend upon the horizontal component of the earth's rate. At a launch latitude of 45 degrees this is about 10 deg/hr. If we assume a gyro drift rate of

0.02 deg/hr such as we did in the earlier section on powered-flight inertial sensing and compute the resulting azimuth alignment error, we get a 2-milliradian error which corresponds to a cross-range miss of 13 kilometers at our 10,000-km target range. This is almost 20 times larger than any miss component previously assumed. The message is clear. If gyrocompass action is used for initial prelaunch azimuth alignment, then the contribution to target miss due to gyro drift will be very much larger than that which that same drift causes during inertial sensing of powered flight.

One mitigating effect can help to reduce this disparity. For a stationary launcher, calibration of the gyroscopes can be accomplished, as will be described later, to correct or compensate for measured

Table 4 Initial Condition Effects on Target Miss.

Error Source (see text)	Standard Deviation Miss at 10,000-km Range Minimum-energy Trajectory Range	Track
Position Error, 100 meters each axis	360	
Velocity Error, 0.1 meter/sec each axis	650	100
Vertical Alignment Error, 20 microradians each axis	260	90
Azimuth Alignment Error, 0.1 millirad		650
RSS Net Effect	790 meters	660 meters

drift. We assume here a residual unknown drift rate of 0.001 deg/hr (0.06 meru) on the launch pad as affecting gyrocompass action which results in 0.1 milliradian (20 sec of arc) of azimuth misalignment. Or alternately we assume any other azimuth alignment scheme which results in 0.1-millirad error. The corresponding cross-range miss at our target is 650 meters.

The relationships and assumptions discussed above on initial condition errors are summarized in Table 4.

Guidance Formulation and Computation Errors

The formulation and in-flight computation of guidance equations will naturally affect weapon performance, but error effects can be kept minimal.

With the type of inertial guidance illustrated earlier in Figure 6, the difficult guidance computations during powered flight are those embodied in the required-velocity computer. An explicit formulation solves in a continuous and exact way the equations of motion as a function of initial position and velocity using a model of the earth's gravitational force field and incorporating other predictable forces and motions through the trajectory back to the bottom of the atmosphere. This results in equations expressing the relationships among the components of the earth-impact location and the components of the required velocity at the present position at present time, assuming rocket thrust termination is commanded. These relations do not define a unique trajectory to a given location of earth impact in the coordinate frame being used. A further operational constraint to select a unique trajectory is designated among the variables of interest. The result is a continuous generation of required velocity components for a unique trajectory with desired characteristics. The mathematical operations can get quite complex and require fast computation speed and large computation capacity.

A designer might, therefore, choose an implicit guidance formulation to reduce the complexity of in-flight computations. Several ingenious techniques have been used but all depend upon flight near a "standard" trajectory for each target range. Following exactly a standard trajectory would require exact knowledge of or active control of the time history of the thrust acceleration from the rocket engine. Throttle control on the engine is possible but can be avoided. The implicit schemes allow variations in the vicinity of the nominal expected trajectory and, for instance, determine required velocity deviations from standard as a power series in the deviation of measured variables from the standard. The wider the variation expected from the neighborhood of the standard trajectory due to predicted statistics of uncontrolled perturbations and the more accuracy in performance desired, then the more higher-order terms needed in the power series. The equation representing the standard trajectory and the power series expressions are computationally simple but obviously require prelaunch determination and insertion of constant coefficients which are a function of the target and launch locations.

The prelaunch determination of the implicit guidance coefficients can be a formidable task in itself, albeit in a ground-based computer. For fixed launch locations and a preselected set of targets to choose from, these can be done laboriously and slowly well ahead of time and stored ready for use. However, if provision is not made to generate new coefficients for new targets quickly, then there is considerable loss in operational flexibility.

Explicit guidance formulations avoid complex prelaunch determination of coefficients at the expense of more complexity in the flight computer. All that the guidance system needs is the identification of the target coordinates and one parameter which selects the particular ballistic trajectory to fly.

Both types of guidance formulation are limited in performance by the accuracy with which the unguided warhead motions following engine thrust termination can be predicted. These effects will be discussed separately. But other than this, both types of formulation can arbitrarily be made more precise by more complexity in the equations solved and in the accuracy of the computer in carrying out the computations. The state of the technology in compact digital computers of abundant capabilities should be able to reduce target-miss contributions from the guidance formulation and computation to negligible magnitudes.

Continuing our assignment of hypothetical miss contributions, we ascribe variances of 100 meters in range and 25 meters in track to the effects of imperfect guidance equations and their solutions.

Thrust-termination Errors

Near thrust termination, the guidance must steer and the thrust-vector control must respond so that all components of the missile velocity will simultaneously become sufficiently equal to the corresponding components of the computed required velocity. When these conditions occur, the guidance system must accurately send a signal to stop the rocket propulsion. The signal must be sent slightly early in recognition that a certain impulse of velocity will occur after the signal is sent to account for the less-than-instantaneous decay of thrust. And the actual thrust-decay impulse must be closely equal to that expected as the engine dies. Accuracy of warhead arrival at the target depends upon the degree to which each of these requirements is met.

Design sophistication and care in the guidance and thrust-vector control systems can reduce the contribution of these target-miss-producing effects to arbitrarily low and trivial levels except for the effects of the variations in the actual thrust decay after thrust termination is commanded.

Liquid-propelled rockets are terminated by stopping fuel to the rocket chamber. Solid-fueled rockets can be stopped, for instance, by suddenly opening large forward-facing ports in the rocket so as to exhaust high pressure gases forward and to reduce chamber pressure suddenly so that the fire burning in the solid grain is extinguished. The sudden expulsion of gases forward will reverse the direction of the net thrust so that if simultaneously the warhead at the nose is detached, the rocket will literally back away from the warhead, leaving it in free fall as desired.

In fact, the manner of physically separating the warhead from the rocket is important. If the whole assembly starts to tumble after thrust termination and before release of the warhead, additional target-miss-producing centripetal accelerations will be imparted. The separation operation should assure a velocity difference between the rocket and warhead so that there is no danger of recontact. The simultaneous release and thrust reversal described above achieves this. Alternately, the warhead can be ejected by a spring. The direction and magnitude of the impulse imparted by the spring must be sufficiently predictable and included in the guidance calculations.

There can be considerable unaccounted-for variability in all these actions without tight control of mechanical, chemical, thermal, and electrical tolerances. At the time of thrust termination, the rocket has been accelerating to levels of as much as 100 meters/sec^2 depending upon target range. After the thrust-termination signal, for instance, the rocket continues to produce thrust as the engine dies down, producing a net velocity impulse which must be predicted and accounted for in the guidance. Variations in this velocity impulse from the predicted average may be large. A 0.1-meter/sec difference between the actual and predicted values, equivalent to stopping the 100 meters/sec^2 in 1 millisecond, causes a 650-meter miss at the target.

Considerable easing in the difficulty of obtaining an accurately predictable acceleration decay at thrust termination, but at the cost of considerable complexity, is achieved by using an additional low-thrust final rocket stage, a so-called vernier stage, to add the last small part of the required velocity. If this stage operated at 1-meter/sec^2 acceleration, for instance, then predicting its decay thrust impulse to less than 0.1 meter/sec is relatively easy. The implementation of a multiple independently targeted reentry system described later can provide an effective vernierlike action from the "bus" in the release of the warheads.

Gravity Anomaly Errors

The science of geodesy helps provide ballistic-missile guidance with the coordinates of the launch and target points and the gravity-field forces in between. We examine the gravity field question first.

The sea-level figure of the earth is represented by its geoid. The geoid is the equipotential gravity surface of mean sea level which is extrapolated across land masses by integrating altitude variations along the plumb-bob vertical. The geoid includes both gravitational attraction and the centrifugal force effect of the earth's rotation, which reduces the radial mass attraction on the equator by 0.35%. The geoid, being influenced by density and topographical variations of the earth, is not a smooth surface. An analytical fit to the geoid such as the classical oblate spheroid or ellipsoid of revolution is uniform and provides compact computational simplicity. If the size and flattening of the ellipsoid (or another analytical shape) could be chosen to fit the geoid perfectly, then equations giving the acceleration of gravity vector as a function of location and altitude could be derived to any desired accuracy. Such equations are needed to predict the motions of a ballistic missile in all phases of its trajectory. But the ellipsoid and geoid will differ enough to cause significant variations in the real trajectory.

The differences between the actual acceleration of gravity vector and that derived analytically for a reference ellipsoid are defined as gravity anomalies.

Observed gravity anomalies as a function of distance along the earth's surface have both rapidly fluctuating and slowly changing components. The slowly changing components have relatively low magnitudes of the order of 0.02 cm/sec^2 at sea level. They can be measured and can be included empirically in the formulas used by missile guidance without undue complexity if accuracy requirements dictate. Accurate worldwide determination of anomaly vectors at missile altitudes is possible by observing the motions of artificial satellites.

The rapidly fluctuating components are usually associated with isolated geological characteristics, mountain ranges, and areas of high tectonic activity. Peak magnitudes occasionally reach upward to 0.7 cm/sec^2 but will usually be of the order of 0.05 to 0.1 cm/sec^2 at sea level. The effect of these at the higher altitudes of a ballistic missile is small. They arise from localized mass variations near the earth's surface. The magnitude of the observed resulting anomalies varies inversely as the square of the distance from these sources.

Since at high altitudes the effects will smooth out and be of smaller magnitude, gravity anomalies of this sort will have their greatest effect on missile motion at low altitudes. Moreover, if unaccounted for, the anomalies have less and less effect as the flight progresses, since the unexpected acceleration has less time to propagate into velocity and position errors before reaching the target. For this reason anomalies in the vicinity of the target can be ignored: those during the first part of the powered flight trajectory must be considered.

In summary, target miss due to gravity anomalies will depend upon the degree to which they have been included in the guidance formulation and will depend particularly upon unexpected occurrences near the launch location. If we assume, conservatively, an equivalent uncompensated anomaly-vector effect of 0.02 cm/sec^2 during the first 100 seconds of launch, then the miss contribution at a 10,000-kilometer range target could be of the order of 50 meters.

Targeting Errors

Before launch the guidance system must be given instructions designating the desired target. The problem reduces to the ability to specify with desired accuracy the target location and altitude in the working coordinate system used by the guidance computations.

A target will be selected by its strategic or military interest and will be identified initially by reference to named structures or geographical features. Coordinates of these features might be available from maps of the area involved, but such maps almost certainly will not give locations in the reference geodetic system desired and will not be of assured accuracy.

Mapping reference systems vary from place to place. This is caused directly by the fact that the true geoid of the earth is complex, with nonsystematic variations resulting from the varied mass distribution of the earth. If the geoid could be actually perfectly represented by a clean analytical function such as one of the reference ellipsoids in use, then every reference bench mark and base leg for each map could be laid out easily in this worldwide system with accuracy limited only by the precision of the instruments used. Among these measurements are astronomical observations relative to the plumb-bob vertical, which hangs perpendicular to the geoid. But the true geoid is not analytical and is known only with various degrees of accuracy around the world. The angle between the plumb-bob vertical on the reference ellipsoid and the normal to the reference ellipsoid being used to approximate the geoid is of interest here. It is called the deflection of the vertical. With careful gravity surveys in the vicinity, the deflection of the vertical can be estimated. Values of the order of 0.1 milliradian (20 seconds of arc) are not uncommon. An unrecognized deflection of the vertical of this magnitude results in an error of astronomical determination of location of 640 meters. Corrections for estimated deflection of the vertical should reduce this significantly.

The undulations of the true geoid relative to the reference ellipsoid also can cause errors in elevation determination with respect to the reference surface. The magnitudes are strongly dependent on the particular choice of reference ellipsoid but are relatively small. Even with ellipsoids chosen to fit the geoid over continental areas, the altitude difference between geoid and ellipsoid is usually within 25 meters. This magnitude of error will occur in altitudes measured from sea-level-referenced surveys unless corrected.

The location of one point astronomically determined in the reference coordinate system as described above can be used as an origin for the mapping system. For triangulation, a single nearby baseline of known length and azimuth is needed. Then the rest of the mapped points can be established geometrically by angle measurement surveys. As the triangulation is carried further from the origin and across difficult terrain, errors can build up to significant magnitudes. Observed errors in closure or repeat measurements indicate that errors of the order of 20 parts per million are possible in careful surveys. This is a 60-meter error in carrying a survey 3,000 kilometers from the origin.

Conventional triangulation survey is impossible across wide stretches of water, and other methods have been used to tie together continental systems with each other with more accuracy than astronomical measurements provide. Recently, artificial-satellite observation has helped enormously the science of worldwide geodesy and cartography.

Obviously, satellites also provide a most useful tool for ballistic-missile target selection, designation, and location. From a 200-kilometer altitude, star-referenced directions to the target of interest measured optically to 0.1 milliradian would locate the target to 20 meters with respect to the satellite at the instant of observation. With the satellite traveling at 7.5 kilometers/sec, timing error of 10 milliseconds would cause another error of 75 meters. As the same satellite passes over reference points located on the reference ellipsoid in use, the orbit can be tied to the ellipsoid with equivalently sized errors. Unknown orbital motions between reference and target observations can be reduced by long-time observation of satellites so as to create accurate earth-gravitation and atmospheric models at orbital altitudes. Unknown anomalies located near the target at orbital altitude will degrade accuracy but one would expect the error to be quite small.

We choose target location uncertainty components of 200 meters for our hypothetical missile system.

Reentry Errors

Without the earth's atmosphere, the warhead would not be deflected from its vacuum free-fall trajectory as it approached the target. But in the real case, strong reentry forces of atmosphere drag and lift radically modify the flight path. The trajectory as it passes through the atmosphere is predicted ahead of time and its effect is included in the guidance formulations of the boost phase. Target-miss errors result to the extent the entering warhead is deflected from the expected trajectory.

The reentering body first senses the atmosphere significantly at an altitude above 50 kilometers as it comes slanting in from space. The first effects are aerodynamic moments which tend to rotate the body to a stable aerodynamic trim orientation. The drag is initially small and the translational motion is dominated by gravity. As the body penetrates deeper into the atmosphere, high drag forces with associated high deceleration and high aerodynamic heating occur. During this period the trajectory slowly tips over from the initial shallow angle of approach toward the vertical. For high-drag reentry-vehicle shapes as the drag reduces the velocity further, the warhead falls almost vertically.

The problems of reentry could be eliminated if it were possible to streamline the shape so that it would slip through the atmosphere unaffected. But the physical laws do not cooperate. Streamline shapes of high aerodynamic coefficient turn out to have high heat generation and heat transfer to the structure. With high-drag blunt shapes the heat generation is higher, but the major part of it is transferred harmlessly to the airstream, as the reentry body is protected by a strong bow shock wave. The penalty of high-drag shapes is early deceleration and a longer time in the lower atmosphere where winds and atmospheric variations can deflect the flight path to cause miss at the target. The design compromise is somewhere between the better trajectory accuracy of low drag shapes and the low heating of high drag shapes.

One of the deflecting forces is the aerodynamic sideways force or lift caused by shape dissymmetry or offset center of mass of the body so that it flies at trim attitude with an angle of attack. If the center of gravity is not within a few millimeters of the axis of symmetry, for instance, significant sideways deflection of the trajectory can occur. Spinning the body about its symmetrical axis will stabilize it and cause the deflecting forces from nonsymmetries to average near zero. Another miss-producing effect is that of nonstandard-atmosphere conditions arising from meteorological phenomena. Variations in air density will modify the drag and shape of the trajectory, predominantly affecting the range component of dispersion. Winds can significantly deflect the trajectory, although the large component of apparent wind due to the air mass being carried eastward by the earth's rotation is predictable and causes no problem.

All trajectory-deflecting effects are helped by designing the reentry body to tolerate the higher heating that low-drag, fast reentries cause. For our hypothetical missile system we choose reentry variations of 300 meters in range and 200 meters in track.

Finally, the warhead reaches the vicinity of the target. If an airburst is desired we may ask about the performance of the fusing system. We will assume that the altitude variations in triggering the explosion are adequately controlled so as not to affect weapon performance significantly.

Overall Missile Performance

Previous sections have identified the major contributors to ballistic-missile target miss. In each case, error components were chosen and related to a 10,000-km-range target on a minimum-energy trajectory. These are summarized in Table 5 and combined together statistically by square root of the sum of the squares (RSS), each contributor being assumed to represent the standard deviation from a zero mean of independent contributors. Range and track components are further combined to a circular error probability (CEP), this being the radius of the circle centered on the target which contains half of the population of miss events of many independent launches.

Table 5 has only little basis in experimental fact. It is primarily educated guesswork and speculation. It represents data on no specific weapon system. Indeed, the reader may have noticed deliberate inconsistencies in assumptions which were made to emphasize important relations.

Table 5 Hypothetical Overall Target-miss Summary (10,000-kilometer Minimum-energy Trajectory).

Error Sources (see text)	Miss and Miss-squared Standard Deviation Components			
	Range		Track	
	(meters)	(meters²)	(meters)	(meters²)
Inertial Sensing	1,050	1,100,000	270	73,000
Initial Conditions	780	610,000	660	436,000
Guidance Formulations and Computation	100	10,000	25	625
Thrust Termination	650	424,000	100	10,000
In-flight Gravity Anomalies	50	2,500	50	2,500
Targeting	200	40,000	200	40,000
Reentry	300	90,000	200	40,000
Sum of Squares		2,276,500		602,125
RSS Net Effect	1,500		770	
CEP (approximate)	0.59 (1,500 + 770) = 1,340 meters			

There was one important objective, however, in the assumptions leading to Table 5. Most current unclassified articles on strategic weapons, stating presumptions on ballistic-missile performance, have used something near one and two kilometers CEP for both Soviet and U.S. missiles. Our round-number assumptions were adjusted with this in mind and we arrived at 1,350 meters CEP.

We do not have to do entirely without any experimental evidence to make the results of Table 5 credible. For instance, the large contributor to miss is the inertial-sensing performance which is most difficult to estimate without citing actual test results. There are several U.S. suppliers of inertial navigators for commercial airliners, all having more or less equivalent advertised performance. There have been tens of thousands of hours of use of these navigators in flights all over the world with extraordinary reliability. Accuracy is usually quoted as 1 nautical mile (2 kilometers) of navigation error per hour of flight. This implies that a level of 0.02 deg/hr or less gyro drift occurs in these airplane navigators operating in the earth's 1-*g* acceleration environment, assuming all the navigator error is attributed to this source alone. The values 0.02 deg/hr and 0.02 deg/hr per *g* were assumed earlier in Table 3 as source data for Table 5. (Such aircraft navigators do not put as great demands on accelerometer performance as does ballistic-missile guidance and cannot, therefore, help substantiate the presumed accelerometer accuracy.)

Perhaps the best experimental evidence to help make the results of Table 5 believable is from data available from spaceflight inertial guidance. A recent report of a series of inertially guided booster launches into low earth orbit, for instance, showed an average error in velocity magnitude at the time of orbital insertion of 0.3 meter/sec and a flight path uncertainty average of 0.007 degree. A boost into earth orbit puts somewhat greater demands on inertial guidance than does ballistic-missile guidance; for a given velocity error at thrust termination, the boost into orbit is slightly harder. However, assuming these values of 0.3 meter/sec and 0.007 degree on our 10,000-km minimum-energy trajectory we see the former causes about 1,800 meters in range miss, and the latter, randomly oriented, causes about 450 meters in range miss and 900 meters in track miss.

Levels and Limits of Guidance Accuracy

Based on evidence available from commercial and spaceflight applications of inertial navigation and guidance, an ICBM CEP of 1 kilometer for present-day weapons is certainly plausible. But actual design intent, error analysis, and test demonstration results by military developers of these weapons are hidden in secrecy. Of equivalent interest, however, in its impact on present and future arms-race and arms-control dynamics is speculation on what theoretical or practical limits to ultimate performance there might be.

Arbitrarily striving to make performance better and better is bad economics. For large warheads designed to devastate large areas such as cities and their populations, delivery to one or two kilometers by a ballistic missile is good enough. Discrete hard targets and smaller warheads, however, require more accuracy. Consider a missile silo hardened to survive an overpressure of 2,000,000 newtons/meter^2 (300 psi in English units) that is attacked by a single small warhead of 20-kiloton yield (Hiroshima size). To obtain a 90% probability that the target will be destroyed, one formula indicates the CEP of this weapon must be 100 meters. If the warhead yield is decreased to 20 tons *or* if the target hardening is improved by a factor of 100, then a CEP of 10 meters is indicated. In this extreme case the formula is suspect, but the trend can be believed.

We will examine briefly each source of error in unaccounted force acting on the missile or error in measuring geometry and motion of the trajectory in an attempt to discover if there are any obvious limits in performance. We look for significant contributions to miss which cannot be controlled or minimized in some practical way.

Several error contributors which can be dismissed easily are those which can be managed by straightforward engineering design. Included are those errors due to guidance-equation formulation and computation. More elaborate and complex equations and more significant bits and speed in the computer solving these equations can match arbitrarily well our understanding of the relationships among the supplied initial conditions, the supplied target coordinates, the measured acceleration, and the required thrust-termination velocity. Also, with a low-thrust last-stage vernier rocket, for instance, with careful control and instrumentation, the actual warhead release velocity can be made to match the computed required velocity to levels within 1 meter target-miss effect. For a stationary and carefully surveyed launcher, initial-position and initial-velocity errors should be reducible to 1 meter effect at the target.

Another class of error sources is those limited by our scientific ignorance. These include gravity anomaly and targeting errors. In the section on gravity anomaly errors, the fine structure of the gravity anomalies at low altitude were estimated to cause a 50-meter error. If these are carefully measured, the effect on a compensated system could reasonably be reduced to less than 10 meters. The high-altitude, slowly varying anomalies can be measured by their effect on satellite motion. Likewise, measurements can be made with straightforward techniques to locate targets with respect to orbiting satellites to within 10 meters. The rest of the problem concerns the degree of complexity used in fitting elaborate mathematical models of orbital motion to careful observation.

Other geophysical phenomena affecting spacecraft and missiles were not discussed in earlier sections but must be examined in these limiting cases. An unexpected acceleration of 6.0×10^{-4} cm/sec^2 acting throughout an ICBM flight of 30 minutes could cause a 10-meter miss. This might arise from an unaccounted-for steady force of 60 dynes acting on a 100-kilogram warhead. Radiation pressure, solar-wind forces, high-altitude atmospheric drag, earth-magnetic-field forces, electrostatic forces, etc. all should be safely below that level. Similarly, movement of the earth's pole and continental drift are inconsequential. The gravitational attraction of the moon and sun need to be considered. At spring tide, the deflection of the vertical due to both moon and sun is a maximum of 0.1 microradian and the anomaly magnitude can be no more than 0.00016 cm/sec^2. These can cause ICBM target miss contributions of a few meters but can be easily compensated for.

The final category of error contributors includes those which are near the limits of engineering and technological state-of-the-art. The inertial sensing measurement errors and the prelaunch alignment are prominent here. At ICBM ranges, a 30-meter CEP, for instance, means measuring and controlling angles to less than 1.5 microradians (0.3 sec of arc) and velocities to less than 0.5 cm/sec. It means gyro acceleration-independent drift rates of less than 0.0015 deg/hr and gyro acceleration-sensitive drift rates of less than 0.001 deg/hr per *g*. It means accelerometer reading offsets at least as small as 0.0015 cm/sec^2 and readout scale factor error smaller than 0.6 part per million.

These levels of better performance should be within the grasp of technology in a new generation of instruments and their application to inertial sensing. New techniques of measurement will be required. For instance, conventional optical transfer of angles to 1.5 microradians could require a prohibitively large optical aperture operating at the diffraction limit. Rather, the instruments themselves will be called upon to sense requisite angles. A vertical-seeking loop using the accelerometers with the above error will put the sensitive axis of the accelerometers themselves within 1.5 microradians of the horizontal. Testing these instruments will require very stable and well-monitored foundations and long periods of observation to indicate error and establish the absolutely essential stability of instrument output. Once stability is demonstrated, the error can be brought to null by compensation such as described in the next section.

Another error source which is limited by the engineering and technological state-of-the-art is that of atmospheric reentry deflection. This can be controlled in either of two ways. First, the reentry vehicle can be designed for a faster, more streamlined shape of high ballistic coefficient at the expense of protection against the consequent heating. With a high ballistic coefficient of 10^4 kg/meter^2 (2,000 lb/ft^2), for instance, the corresponding atmospheric-density and wind-caused dispersions are of the order of 70 meters. If further improvement by increasing the ballistic coefficient does not seem feasible, then the warhead can be inertially guided through reentry to follow a precise trajectory in a manner described in a later section.

Based upon these considerations, this writer estimates an overall ICBM CEP of 30 meters may be expected with reasonable and practical application of science and technology to the task. Furthermore, this writer believes that an order of magnitude margin must exist before the limits of ultimate dimensional stability of known materials can be considered. Such performance is possible. It will become reality if the government and military incentive cause resources to be committed to the challenge.

The Art of Gyros and Accelerometers

Achievement of high-performance, quality gyros and accelerometers such as now in use for inertial navigation has been the result of many years of effort by many teams of engineers and scientists. It has been pioneering technology into areas not fully understood. Every participant can remember brilliant ideas which turned out to withhold the elusive promise. Every participant can tell stories of frustration when a technique or design which showed excellent results in test suddenly and uncontrollably went bad with no apparent explanation. The design, development, and manufacture of inertial sensing components is an unforgiving art.

Why is this? A gyro with a wheel angular momentum of 10^6 CGS units is a practical size. A drift of 0.01 deg/hr will be caused by uncontrolled torques, such as bearing friction on the wheel support, of only 0.05 dyne-cm. Or this could be caused by the earth-surface weight of a spurious piece of dust of only 0.05 microgram attached 1 centimeter from the torque-summing axis. Or it can mean a shift of the center of gravity of a 500-gram spinning wheel with the above momentum, with respect to the support structure torque-summing axis, of only 10^{-7} cm or 10 angstroms. This is remarkable geometric stability for a motor-driven high-speed rotating member in high-stiffness bearings expected to last for hundreds or thousands of hours before ultimate end of use.

Increasing wheel speed to increase angular momentum so that error torques will cause less gyro drift can require so much more wheel motor power and associated thermal problems that the center of gravity shift caused by temperature gradient can rise faster than the angular momentum so as to result in poorer performance.

The problems in accelerometers of inertial-guidance quality are equivalently difficult. Ballistic-missile inertial sensing must measure specific force levels; for instance, from 0.01 cm/sec^2 to 100 meter/sec^2. This means measuring the force supporting the accelerometer test mass over a dynamic range of 10^6 with a linearity and calibration of perhaps 10 parts per million. Typical force-measurement techniques such as strain gauges or calibrated beams fall far short of this accuracy and range. In a pendulous configuration of the test mass, a nonsaturating and absolutely linear restraint can be derived from angular acceleration of an inertia or angular velocity of a gyro-wheel angular-momentum vector. The latter is the technique of the pendulous integrating gyro accelerometer. A balance is achieved within the instrument from the same internal structure torque which accelerates the pendulous test mass and precesses the spinning gyro wheel at an angular velocity which becomes proportional to the sensed acceleration. Angle reading with a suitable electromechanical transducer must cover the 10^6 dynamic range and 10^5 linearity mentioned above. Miscellaneous spurious torques of minuscule size and near infinitesmal changes in structural geometry cause error.

The art of gyros and accelerometers has pushed developments in precision machining and measurement; it has forced new knowledge in the dimensional stability of materials; it has demanded unusual new materials with special properties; it has developed new bearings of very low and zero friction and high-speed bearings of enormous stiffness, stability, and life; it caused the development and refinement of many new electromagnetic transducers of phenomenal accuracy; it has required highly specialized assembly, calibration, and test techniques in special clean-room facilities.

It is interesting to note that the equivalent gyro drift of the ship gyrocompasses early in this century and the acceleration resolving power of the gravimeters of the late nineteenth century meet those same requirements as needed for ballistic-missile guidance. These pioneering instruments were hardly configured for ballistic-missile guidance and the rocket environment. But even so, obtaining precision inertial-sensing performance in small rugged instruments is relatively easy for short periods of time. The trick is to construct an instrument which will give constant high-quality, uniform results over periods of time long enough to be useful.

The most performance can be wrung out of an inertial-sensing instrument by calibrating it as close as possible in time to the period of use. Calibration means the instrument is tested and either adjusted to near perfection or the indicated error is measured carefully and a compensation applied in the computations which use the instrument output. This calibration can take place in the missile just before launch by programming various attitudes of the sensor stable platform to various orientations with respect to the earth's rotation vector and the earth's gravity-force vector. With earth rotation and gravity as test inputs the needed calibration of or compensation for most of the various component instrument outputs can be determined. This works only with a stationary launcher, of course, but is a process which can proceed continuously up to the signal for launch. Limited gyro wheel-bearing life might contradict such continuous operation and calibration, and if instrument stability is not built in, launch might have to wait until calibration can be accomplished to get best performance. But then, a quickly responding "second strike" retaliation may not need best performance.

We can tolerate perhaps 0.03 deg/hr of gyro drift if we wish our missile to land within 1,000 meters at 10,000 kilometers. Having that level of drift in the test laboratory doesn't necessarily do the trick. Back in 1924, 0.001 deg/hr was demonstrated in the laboratory. Transfer of laboratory results to field operation is not simple. Predictable and consistent performance is needed while in flight and while subjected to the rough vibrations and high acceleration of rocket burning.

The focus of most effort in inertial-instrument design is in stability—stability in materials from chemical changes, stability in materials from physical changes arising from unrelieved strain, and stability in dimensions and other properties by careful and continuous precision control of temperature, temperature gradient, and electrical input power.

The art of gyros and accelerometers has been a most difficult but a most successful one.

Launch from Moving Bases

Most of the previous discussion has assumed a fixed base or at least a stationary launcher. Deploying your missiles in moving or movable launchers is of enormous advantage in discouraging a first strike against you. But just as your enemy will have difficulty in locating you as a target, you may have difficulty in finding out precisely where you are launching from. The problem of guidance of ballistic missiles launched from moving or movable bases is in the determination of initial conditions.

The accurate launch from submerged submarines cruising beneath the sea out of sight of landmarks is conceptually remarkable. The missile guidance needs launch latitude, longitude, and altitude (depth); it needs the directions of vertical and north; and it needs all three components of present velocity. Again techniques using inertial sensing supply the answer in the form of inertial navigation of the submarine.

The following description is oversimplified, but it does suggest what is involved. These submarine inertial navigators depend on comparing the direction of the accelerometer-sensed local vertical with a gyroscopic physical memory of a rotating-earth reference coordinate frame. The direction of the vertical in this frame locates the submarine. The direction of north is discovered by gyrocompass action as described in more detail in an earlier section. Velocity components at the location of the submarine navigator are obtained by processing accelerometer signals and can then be extended to missile-guidance sensor location by adding terms equal to the angular velocity of the submarine times the distance between the two points.

Inertial navigation of the submarine looks at first very different in concept from that for the missile described earlier. The submarine navigation is that of low acceleration over a long time period; that for the missile is of high acceleration over a short time. But there is an underlying basic similarity. The diagram for missile inertial navigation in Figure 5 applies. The gravity-computation feedback from position back to acceleration surrounding the two integrations, if disturbed with error, will oscillate with a long period, that is, the 84-minute Schuler period. This does not appear in the short times of missile powered flight of 5 minutes or so but is prominent in the submarine navigation. Error oscillations of both the Schuler period and a 24-hour period will build to higher and higher magnitudes due to imperfect behavior of the gyros, accelerometers, and other sources of error. The submarine inertial navigation will need periodic initialization.

If the submarine sits stationary on the bottom of the ocean, then with suitable accuracy of the inertial sensors the vertical, north, latitude, and all velocity components can be determined initially without any external data by inertially sensing the gravity-force and earth-rotation vectors. There is no way to get longitude by inertial means. Indeed, the prime meridian locating zero longitude is arbitrarily defined through Greenwich. The submarine inertial navigator needs initial data and, if in motion not accurately measured by other sensors, will need them for other variables as well as longitude.

How long the submarine can go without an external position fix of latitude, longitude and direction is a function of the quality of the inertial sensors. An uncompensated gyro drift of 0.002 deg/hr causes navigation error amplitude to build up at a rate of 220 meters for every hour without a position fix. The weapon-system designers desire as long a period between position fixes as possible, since position fixes are obtained only with some increased jeopardy of exposure to enemy detection. The ocean is an excellent hiding place except when peeking out to find out where you are.

Multiple Warhead Guidance

By using a single rocket to launch a number of independent warheads, the probability of penetration of a missile defense system is greatly improved, although the total warhead yield will have to be reduced considerably to account for the added weight of the extra systems for warhead release, reentry, arming and fusing, and thermonuclear-warhead triggering.

A multiple warhead release by springs or by small fixed solid propellant rockets, for instance, can provide an array of small velocity increments to the reentry vehicles in various directions to cause the several warheads to land in a fixed pattern around the aim point. This has been called multiple reentry vehicle (MRV).

Rather than a fixed pattern around a single aim point, more sophisticated arrangements can adjust the individual velocity increments so that each reentry vehicle is caused to follow a path to individually selected targets. This has been called multiple independently targeted reentry vehicle (MIRV).

There are a number of different possible MIRV configurations. If the multiple warheads are intended to destroy a number of discrete targets arrayed in a pattern, such as a field of enemy missile launch silos, then the springs or small rockets providing the velocity increments can be preadjusted before launch in both direction and magnitude so as to match that particular target pattern.

Another method of implementing a MIRV would provide each warhead with its own small rocket and with an individual control and guidance system. Each would then be steered separately through the velocity increment appropriate for the target specified for it by the instructions loaded into its guidance system before launch.

Still another configuration provides a low-thrust final stage, a so-called bus, which carries the single guidance system and all of the reentry vehicles with provision for them to be released one at a time. This low-thrust bus is then guided through a sequence of velocity changes. After each velocity change, one of the warheads is released from the bus toward the target defined by the velocity change achieved at that time.

For each of these configurations, the maximum spread of targets which can be covered in a single launch depends upon the magnitude and direction variations of the velocity increments which can be provided by the springs or small rockets. Suppose this velocity change which can be added is 20 meters/sec maximum for a particular construction. At the minimum-energy ballistic velocity for a 10,000-kilometer range, 20-meters/sec velocity increment in the appropriate direction can change the target impact range by about 120 kilometers, but the same increment to the side can change the target impact in track by only 20 kilometers. The maximum dimensions in target impact spread of MIRV systems will always be much longer in the range direction than is the width in the track direction. The size of such patterns of reachable targets can be increased by providing for more velocity-increment capability, but the payload weight used for this leaves less for other things such as the number of reentry vehicles or less for the warhead yield of each reentry vehicle.

If a ballistic missile which is designed to send a single large warhead to a single target is converted to MIRV capability, what degradation in accuracy of the smaller warheads at their targets can be expected? We can assume that the major part of the ballistic velocity is applied in each case with equal accuracy. The question then concerns only the accuracy with which the velocity increments are applied to the individual warheads. For a 10,000-kilometer range the velocity already achieved is 7,000 meters/sec. The added velocity will be very much smaller. A given percentage-magnitude error or a given direction error in applying this velocity change will have a relatively small effect on target miss.

There is no theoretical justification to feel that individual MIRV warheads will not have a CEP essentially the same as that of a single-warhead missile using the same level of technology.

In-flight Star Tracking to Improve Accuracy

The accuracy of an inertially guided ballistic missile is dominated by the errors that have accumulated during the prelaunch and early phases of powered flight. If the accumulated error could be reduced before the end of powered flight by some new measurement process, then the guidance could steer to a more accurate thrust-termination velocity with accompanying reduction of miss at the target. Theoretically, the three vector components of position and three vector components of velocity would all have to be measured somehow and used to correct these values in the inertial navigation to do a complete correction for accumulated error. Not all six values would necessarily need to be measured, however, to cause significant improvement since they do not have equivalent sensitivity in effect on target miss. The requisite measurements could be performed by cooperative ground radar or radio, but such would severely compromise the strategic invulnerability of the system.

An independent source of data which in some situations would be useful in improving CEP is the making of onboard sightings of celestial objects late in powered flight when the rocket has climbed up above the atmosphere and obscuring clouds. Such sightings cannot be compared to astronomical navigation sightings made near the surface of the earth on ships and airplanes. The latter depend upon making navigation angle measurements between celestial objects and the local horizon sensed either by a pendulous sensor or optically to the visual horizon.

On the accelerating missile a pendulum will not indicate the local horizon, and the physical horizon cannot be sensed accurately enough through the scattering and emitting atmosphere at any wavelength to improve navigation accuracy.

A practical onboard celestial-object sighting which can be made will measure the angle components between an identified star of known celestial coordinates and the inertial-sensing stable platform. With no error, the measured direction of the star relative to the platform would equal a precomputed value that depends directly and only upon the intended celestial orientation of the platform and the celestial coordinates of the star. If the measurement differs from that expected, the platform orientation is in error by the indicated amount. There can be a number of sources for this error. Clearly gyro drift from launch to the time of measurement is involved. But so is the initial platform alignment which, let us say, is intended to be oriented to the indicated local vertical and some azimuth from indicated north at the time and place of launch. If the clock used by the launch process is in error, then the celestial coordinates of the vertical differ from that expected, since the local vertical is rotating at a component of earth's angular velocity of 0.07 milliradian per second. If the actual launch location in latitude and longitude differs from that indicated, then again the platform orientation in celestial coordinates will differ from that expected by a magnitude equal to the geocentric angular error in launch location.

So a difference observed during powered flight between the measured and expected star direction with respect to the stable platform could be due to any one or more of several error sources. To make use of the measured star-direction difference for error compensation purposes, the error source must be identified or presumed since the effect of each on target miss can differ substantially.

A single star-direction sighting will give only two components of data. Sighting on a second star at a substantial angle from the first star can give added data, but in a practical situation only one star properly chosen should provide nearly the full value that star sightings can give.

If the weapon-system designer concludes that initial platform-alignment errors will be sufficiently small, he may help his inadequate inertial sensing gyros by a single star sighting, choosing an acceptable star as near in line as possible with the expected velocity vector at thrust termination. Rotation error about this direction has little effect on miss at the target.

If the weapon-system designer concludes that launch location errors will be relatively large, such as might be the case for a movable or moving launcher, then he might be able to improve CEP by choosing an acceptable star as nearly directly above the target at impact as possible.

Help from this technique is of value only when there is a clearly dominant source of error which affects the celestial orientation of the platform at the time of star sighting. The technique can do nothing about errors in the performance of the accelerometers, for instance.

In use, once the measurement is made, the guidance computations would modify the steering for the remainder of powered flight to reach a modified, and presumably more accurate, thrust-termination velocity vector.

Fractional-orbit Bombardment

We discussed briefly in an earlier section the characteristics of ballistic trajectories which spanned geocentric range angles greater than 180 degrees. Effecting a missile attack by going the long way around to the target can afford an element of surprise by approaching from an unexpected direction. This is one form of so-called fractional-orbit bombardment (FOB). These trajectories have excellent features for geocentric range angles less than 300 deg, but when called upon for greater angles they need high velocity from the launch rocket, make markedly increased demands on guidance performance to achieve a given CEP, and arrive at very shallow reentry approach angles which also tend to degrade accuracy.

Another form of FOB can relieve some of these difficulties for very long ranges and add other attractions for the weapons designer. In this form the booster rocket is guided to put the missile into a low earth orbit just above the atmosphere and in the appropriate plane. A few hundred kilometers short of passing over the target, a second rocket thrusting period is initiated and the reentry body is guided in a de-orbit maneuver to get it into a descending trajectory back through the atmosphere to the target. This type of FOB has the advantage that until the de-orbit maneuver has been performed, there is little that the defense can do by watching the trajectory to predict what target is being attacked or even if the orbiting body is aggressive. With some cost of added propulsion, a considerable out-of-plane maneuver at de-orbit is possible for attacking a target not in the orbital plane.

The weapon system designer can choose the de-orbit maneuver so as to descend at either a shallow or steep angle. A shallow descent requires considerably less propellant in the de-orbit rocket and puts only small demands upon the accuracy of guiding the maneuver. This latter benefit may be lost in the increased atmospheric deflection accompanying a shallow atmospheric reentry. Other than the surprise element cited above, there is perhaps little advantage in a shallow reentry FOB, since except for extreme ranges an all-ballistic free-fall trajectory from the booster requires less propulsion and is more accurate.

A steep reentry FOB minimizes the reentry deflection problem. Moreover, the same steepness of entry in a conventional trajectory from boost puts the incoming warhead above the horizon as seen from the target area and in sight of defensive radars for a long span of time. The steeply entering FOB will stay below the horizon until it gets close and then, of course, does not announce its intended target or terminal trajectory until its de-orbit maneuver a minute or two before it reaches the earth's surface and explodes.

Obtaining a steep de-orbit and reentry can be very expensive in de-orbit propulsion. For 30 degrees from the horizontal about 4,000-kilometers/sec velocity change is needed, which is comparable to the 8,000 kilometers/sec needed from the launch rocket to get into orbit in the first place.

The guidance of these de-orbit maneuvers is not particularly difficult nor particularly demanding of inertial-guidance performance. The inertial-sensor alignment error at the start of the de-orbit maneuver is degraded by all of the gyro drift occurring during the boost and free-fall phases. An error in achieving the desired velocity change, however, does not propagate into as large a target miss as would the same error in the initial boost into orbit or into a conventional trajectory because of the much shorter time and shorter distance before the target is reached.

A given inertial-guidance system might have twice the CEP in FOB use that the same system would provide with the more conventional ballistic trajectory.

If one considers a multiple-orbit bombardment system (MOB ?), now prohibited by treaty, the question of accuracy becomes very significant. If the attacker delays the de-orbit until he has had opportunity to track the missile one or more passes over his territory, he can establish a very precise knowledge of the orbit and can therefore radio extremely accurate de-orbit instructions up to his missile.
A given inertial-guidance system could provide several times smaller CEP than the same system would produce with the conventional ballistic trajectory. This feature has unfortunate destabilizing first-strike implications. Moreover, such a system could be tested thoroughly in secret by returning dummy warheads onto home territory such that the whole

operation would look like a legitimate nonaggressive operation.

Controlled and Guided Reentry

Active control of the flight path of a reentering warhead might be provided either to correct the deflections away from the expected trajectory caused by atmospheric variations or to provide for a terminal maneuver to confuse the defense. Moving control surfaces or an offset center of gravity with accompanying lift at the aerodynamic trim orientation can supply deflecting forces to be commanded up, down, left, or right as required. The commands would be generated by a guidance system which almost certainly would be based on inertial sensing.

An inertial-guidance system in the reentering warhead must tolerate the large-deceleration environment forces without failure or significant performance degradation. More performance error can be accepted in the inertially sensed acceleration magnitude and direction than is required during the boost guidance phase for equivalent miss at the target. This is because the time of flight in the atmosphere is so short. Even a low-ballistic-coefficient, high-drag warhead will be in the atmosphere not much longer than one or two minutes before impact. Accelerometer error as high as 1 cm/sec^2 will cause only 100 meters or so miss. The inertial system stable-platform alignment at the beginning of reentry must be considered. Gyro drift of 0.02 deg/hr since the initial prelaunch alignment will result in only 0.01-deg platform alignment error 30 minutes later. This could produce as much as 50-meters miss in a guided reentry. But the same 0.02-deg/hr drift during the inertially guided launch phase causes perhaps ten times this miss contribution.

The inertial-sensing performance of these guided entries is dominated by the effect of the high acceleration which brings higher-order nonlinearities into prominence. The inertial sensors must be designed with higher mechanical stiffness and symmetry in anticipation of the high structural loading.

These facts apply primarily to straight-in reentries steered with a low lift in a deflection-minimizing path so as to reduce target miss arising from air-density profile and wind variations. To obtain the considerable maneuvering needed to confuse the defense, a rather high lift is required. In this case, design might be directed toward a high lift-to-drag ratio glide vehicle with a long period of high-speed, near-horizontal flight in the atmosphere. Such a vehicle would come in low where defense radar would have difficulty seeing it, and its considerable maneuvering capability would make its future trajectory and ultimate target impossible to predict.

The guidance of such a glide-entry vehicle would be somewhat more difficult for a given accuracy than guiding a conventional straight-in fast entry. But with such a configuration, terminal sensing and guidance become practical. With radiation sensing of the earth below, terrain matching to stored patterns of the local features in the target area can be used for guidance data. With good inertial-guidance accuracy in the vicinity of the target, the terrain-matching process need not be particularly complex since the initial error is small and the orientation of arrival to the target is relatively fixed or predictable. The final maneuver is to steer the warhead to the sensed target with pinpoint accuracy.

Conclusion and Comment

The long-range ballistic missile, a wonder of modern engineering, is a terrifying weapon. It can deliver a thermonuclear warhead of unprecedented destructive power right on a target 30 minutes and a continent away. Two great nations of the world have built up enormous arsenals of these weapons targeted against each other in an arms race of response and counter-response. Each side attempts to gain some competitive advantage or, by threatening massive revenge, to deter the other from striking first. Other nations have felt compelled to join this race. Defensive anti-ballistic-missile systems are beginning to appear in effective configurations and quantities. To penetrate these defenses, more complex offensive weapons in the form of multiple-warhead missiles have been deployed, with multiple independently targeted configurations nearing deployment. Fractional-orbit bombardment forms may also be appearing and maneuverable reentry looms ahead as an option for future escalation.

In this environment, the two involved nations are meeting with cautious hope in search of workable arms-limitation treaties. Observable and therefore enforceable treaty provisions and agreements are urgently needed to terminate the arms race.

The dynamics and control of strategic arms is an inexact, controversial science. I have chosen in this concluding section to examine some of the possible implications of ballistic-missile guidance accuracy as applied to strategic arms interaction with or without arms limitation treaties.

Using the scanty evidence available, I have estimated that existing weapons are at least as accurate as 1,000 meters CEP. I further estimated that present technology is easily capable of achieving far better accuracy. Miss at the target of 30 meters or less CEP at ICBM ranges can be developed.

In spite of the arms-race escalation, many feel that the retaliatory threat of ballistic missiles to cities has been the success factor in preventing major war. Higher accuracy than currently available in ballistic-missile guidance is largely wasted on "city busters" which have provided this stabilizing deterrent. Why, then, consider the development of greater accuracy in ballistic missiles?

It is a curious paradox that this peace, such as it is, seems to be maintained by the deterring threat of opposing missiles targeted to cities and populations, and yet no actual launch against these targets has any logic to it. Arguments that a threat against opposing populations must be carried out if its deterrent effect has failed just to maintain credibility cannot stand analysis. Credibility in the eyes of what witnesses? On the other hand you might launch against cities in revenge; certainly not impossible, but clearly an irrational action.

There is no compelling reason to destroy cities and populations. There are many valid reasons against such action. The cities pose you no immediate threat. You would do better to save your resources or expend this ammunition against any remaining real threat. And if that threat has been removed or is unknown, it is still to your advantage to spare the cities for their potential value as viable targets. The threat over those cities might still act as a deterrent against further aggression or could enhance your bargaining position. Furthermore, if you feel that you have won the exchange in some sense, you should want to preserve these cities for their value in postwar recovery.

Why, then, is the present deterrent stalemate of hostage cities effective? It is because you cannot depend upon rational behavior by your adversary. Weapons of both sides are still presumably targeted to cities so as to emphasize the threat. The pessimistic conservative evaluations of the enemy that have fed the arms race has also helped prevent the use of these same arms.

Recognizing that the threat of massive population destruction is useful in maintaining the peace, but actual launch against cities would not be useful, we cannot expect strategic-weapon-system planners not to pay attention to the military capability of their weapons. Military capability of ballistic missiles is improved by more accuracy.

The development of high accuracy in ballistic missiles may be interpreted as an intent to achieve a first-strike capability by which an aggressor, by firing first, can destroy his adversary's weapons and ability to respond in retaliation. The development of high accuracy coupled with a multiple-warhead capability may appear even more sinister. Each MIRV missile of suitable accuracy that is launched can theoretically destroy several enemy missiles in their launchers. By firing first, a distinct advantage in exchange ratio is enjoyed by the aggressor. The enemy's force is destroyed faster than his own is consumed.

Fortunately, the mobile launchers and particularly the strategic missile firing submarines on both sides remove the attractiveness of a first strike. Targeting against a movable missile launcher is extremely difficult and uncertain, and the submarines just can't be seen.

Attempts at international agreement to limit accuracy of ballistic missiles would be futile. Unlike many other manifestations of weapon-systems development, the improvement of accuracy cannot be observed by the other side if such development is intended to be secret. The testing of a high-accuracy ballistic missile can be made to appear no different than an ordinary test unless it suits the testing nation's purposes, since target identification need not be announced. In such a situation, each side may suspect that the other is developing high accuracy and therefore proceed with its own development so as to maintain parity. More accurate systems are almost certainly coming, and when they do come we may not know for sure that they have arrived.

Military considerations alone will not be needed to bring accuracy improvement. The major contributor to missile error is inertial-sensing performance and this technology will be improved in any event for nonmilitary purposes. The enroute and terminal navigation of commercial airplanes, all-weather aircraft landing, air-traffic control, spacecraft guidance systems, and research and commercial submarines all benefit from the self-contained, continuous, wide-bandwidth, accurate data that inertial sensing can provide.

Even though much-improved ballistic-missile guidance is probably inevitable, it need not be justified on grounds of first-strike capability. High accuracy to knock out enemy missile silos can also be justified for a damage-limiting counterforce responsive strike. As I have said, if for any reason of folly the ballistic-missile exchange is underway, there is no rational incentive for either side to aim at populations. Rather, each side should and presumably would try to neutralize the other's remaining force. And it would be best for the rest of the world if such an exchange of weapon against weapon were a precision affair.

Against discrete military targets, highly accurate guidance should allow very much smaller yield warheads to be used since the damage radius varies as the cube root of the yield. A small, accurately guided warhead could be characterized as being neat. It is the humane alternative to the massive city-destroying weapons which even if deployed against military targets can devastate large surrounding areas of no military importance and can poison large volumes of the atmosphere and large areas of the earth with radioactivity.

There will be a practical limit, however, to how small a reentry warhead system can be made. Moreover, as the size and weight of the reentry system gets smaller, the deflection effects during the reentry passage through the atmosphere will begin to grow to the extent that they will dominate and limit overall accuracy, unless the reentry is actively guided. Nevertheless, it is safe to say that highly accurate reentry warheads could be made considerably smaller in size and yield than the presently deployed configurations.

Strategic-weapon-system planners will not give up unilaterally the city-threatening, assured-destruction, deterrent capability of their weapon force. With rocket payloads separated into smaller independently guided warheads of individual low yield instead of the large single warhead of greater total yield, the penetration of defenses is significantly improved. Moreover, the destructive capability against cities in an accurately distributed warhead pattern and the deterring threat may not be compromised.

Small, accurate, ballistic-missile warheads have a clear military function. With these available, discrete targets can be engaged without needlessly killing people. Among these targets could be military installations, industrial facilities, dams, bridges, power plants, and transportation and communication centers. When would such targets be engaged? The possibility of limited or slowly escalating nuclear war must be considered. What should be your response if a single missile or a few missiles are launched against you either by accident or by a politically calculated forceful challenge? Further, suppose these do not land in cities and deaths are relatively few. If your arsenal was designed for deterrent only, was the provocation great enough to destroy an enemy city? Probably not. The enemy may have gambled on this in his deliberately limited first strike. If his launch was an accident he doesn't deserve the ultimate punishment.

Small-warhead accurate missiles in your launchers would provide the flexibility for a selective but credible response. The preannounced surgical removal of a few enemy military or industrial installations with the explicit and obvious intent to minimize loss of life might be a clear and emphatic message of your power and restraint. It could terminate the war.

In summary, if each side has a nontargetable deterrent force, as is the case in the current standoff, highly accurate missile guidance is not destabilizing. Without some changes in international relations, highly accurate guidance should be expected. It alone should not and cannot be prevented by treaty. As it arrives, however, it could allow a dramatic reduction in the size of individual warheads. If war starts, it could allow the humane alternative to deliberate actual strike against cities but still preserve the deterring threat. It could give the military and political leaders more choice of action, more selectivity in choice of discrete military targets, more freedom and incentive to avoid massive killing, and more possibilities for war-termination action.

References

1. R. H. Battin, *Astronautical Guidance,* McGraw-Hill, New York, 1964.
2. C. Broxmeyer, *Inertial Navigation Systems,* McGraw-Hill, New York, 1964.
3. C. S. Draper, W. Wrigley, and J. Hovorka, *Inertial Guidance,* Pergamon Press, New York, 1960.
4. W. Haeussermann, "Description and Performance of the Saturn Launch Vehicle's Navigation, Guidance, and Control System," Third IFAC Meeting on Automatic Control in Space, Toulouse, March 1970.
5. W. A. Heiskanen and F. A. Vening Meinesz, *The Earth and its Gravity Field*, McGraw-Hill, New York, 1958.
6. S. Lees, ed., *Air, Space, and Instruments,* McGraw-Hill, New York, 1963.
7. C. S. O'Donnell, ed., *Inertial Navigation, Analysis and Design*, McGraw-Hill, New York, 1964.
8. G. R. Pitman, Jr., *et al.*, *Inertial Guidance*, John Wiley & Sons, New York, 1962.
9. M. B. Trageser and J. M. Dahlen, *Guidance Characteristics of Ballistic Trajectories of Range Angles of More than 180°*, M.I.T. Instrumentation Laboratory Report E-915, August 1960.
10. A. D. Wheelon, "Free Flight of a Ballistic Missile," *American Rocket Society Journal*, September 1959.

Summary of Discussion

High-accuracy Guidance. Because of the large number of uncertainties, it is difficult to estimate the length of time required to achieve the projected 30-meter accuracy. However, with funds of the magnitude of those expended in the Apollo program, such a system could probably be developed and available for deployment by about 1980. Even without the deliberate application of funds for this purpose by the military, the ongoing research to meet commercial and space needs and the accompanying gain in experience will continue the trend to better accuracy. It is thought that the ultimate limitation on missile accuracy will not arise from the inertial sensing but rather from the scientific ignorance of the earth's gravitational field. Computers by no means represent any limitation.

Unilateral Verification of High-accuracy Testing. An important and controversial question is whether an alien observer in the vicinity of the splash point of a missile test could determine whether the test was of a very-high-accuracy reentry vehicle. If it were possible to be confident of one's unilateral capability in this regard, new areas of treaty agreement might emerge. It was suggested that either of two methods of handling atmospheric dispersion might give away a high-accuracy test. If the reentry vehicle had a very high ballistic coefficient, this might indicate an attempt at very high accuracy. The alien observer may be able to estimate unilaterally that ballistic coefficient. Second, the observer is able to have an accurate profile of wind and weather conditions at the area of reentry and could therefore calculate the expected trajectory of the reentry vehicle through the atmosphere. If the vehicle did not follow that trajectory, the observer would know that terminal maneuvering took place. In order to actually calculate the expected trajectory, detailed information on the six degrees of freedom and the shape of the reentry vehicle would be needed. In many cases it could appear to have terminal guidance but not really have it. A steep angle of reentry would also suggest a high-accuracy test. In general, the absence of high ballistic coefficient and terminal maneuvering would probably indicate that very high accuracy was not being attempted. But to make the opposite judgment would be much more difficult. It was also suggested that an observer might be able to tell from a knowledge of the instrumentation of the experimenter's ships what degree of accuracy he is expecting to measure. It was observed, however, that this would not be a reliable means of judging missile accuracy.

On the Possible Military Significance of the Superheavy Elements

P. L. Ølgaard

During recent years theoretical investigations of the stability of heavy atomic nuclei have indicated that nuclei with masses and atomic numbers well beyond those of the presently known heavy nuclei should be stable enough to allow detection and identification, provided they can be produced in one way or another. These nuclei are called the superheavy nuclei, and the basis for the prediction of their existence is that calculations have shown the presence of closed shells in the region above the presently known nuclei. Had these closed shells not existed, all of the superheavy nuclei or elements would decay by spontaneous fission with extremely short half lives.

It has been predicted that the superheavy elements are fissionable upon neutron capture, and that the number of neutrons emitted per fission, ν, should be as high as 10, while ν for the ordinary fissionable nuclei such as U^{235} or Pu^{239} is only 2.5 to 3. This means that if sufficient amounts of these elements can be produced, it should be possible to manufacture nuclear weapons with weights and sizes significantly lower than those of present-day nuclear weapons. Such a possibility is not without military interest, even if it is very remote.

In order to attempt an evaluation of the military

Dr. P. L. Ølgaard is a Nuclear Physicist in a Senior Staff position at the Research Establishment Risø of the Danish Atomic Energy Commission.

significance of the superheavy elements it is proper to raise and to try to answer a number of questions, the first of which is: Do the superheavy elements really exist?

So far the evidence of the existence of the superheavy elements has come from theoretical investigations. Calculations have shown that there seem to exist regions—sometimes referred to as islands of stability—of almost-stable nuclei well beyond the presently known elements. It seems to be generally agreed that one such region or island of stability is composed of nuclei with atomic number Z around 114 and neutron number N around 184. There may well be other islands of stability, e.g., around $Z = 164$ and $N = 318$, but we will in the following limit ourselves to the region around the doubly magic nucleus ${}_{114}X^{298}$.

A number of attempts have been made to verify experimentally the existence of the superheavy elements, but so far without positive result. These investigations have included measurements on cosmic rays, meteorites, and terrestrial materials, as well as on accelerator-produced nuclear reactions. The investigations have, however, been carried out under conditions where a positive result was by no means certain even if the superheavy nuclei do exist, and consequently the results obtained so far cannot be taken as an indication of the nonexistence of the superheavy elements. Nucleosynthesis by the use of nuclear accelerators would be a very direct way of proving the existence of the superheavy nuclei.

Unfortunately present-day accelerators cannot produce such reactions, but there is little doubt that realistic experiments can and will be performed in a not-too-distant future.

In answer to our first question we can therefore say that there is a reasonable chance that almost-stable, superheavy elements do exist. If we are to continue our deliberations we have to accept this conclusion anyway.

The next question to be asked is: Do the superheavy elements have half lives long enough to make them suitable for weapon purposes?

For detection of the superheavy nuclei it is sufficient that they have half lives of less than a fraction of a second. Obviously much longer half lives are needed for weapon purposes. It seems reasonable to assume that the total half life of the relevant superheavy isotopes must be of the order of ten years or more, while a half life of the order of one year would be too short, both for economic and operational reasons. In this connection it may be remembered that tritium, used in thermonuclear weapons, has a half life of 12.5 years.

The superheavy elements around ${}_{114}X^{298}$ are, as mentioned above, quite stable against spontaneous fission, but the relevant nuclei must of course also be reasonably stable against alpha and beta decay.

Various predictions of total half lives of the superheavy isotopes around ${}_{114}X^{298}$ have been made. While they differ considerably in predicted half-life values, they usually come out with several isotopes (not necessarily the same) which have total half-life values of more than ten years. Sometimes the predicted total half lives are orders of magnitude higher.

It is not sufficient that the total half life is more than ten years for relevant isotopes. It is well known that one of the difficulties in the production of plutonium weapons is the relatively short half life of Pu^{240} towards decay by spontaneous fission, 1.3×10^{11} yr. This means that the higher the Pu^{240}-concentration in the plutonium weapon, the faster must be the compression during the implosion process if the desired explosion yield is to be attained. If we assume that the spontaneous-fission half life has to be longer than 10^{11} years, and that the total half life has to be longer than 10 years, we still find from the theoretical half-life predictions that there should exist some, perhaps only a few, isotopes in the region around ${}_{114}X^{298}$ which fulfill these requirements. The corresponding elements are generally predicted to have Z-values between 110 and 114.

Hence our answer to the second question is that quite likely some of the superheavy isotopes have half lives long enough to be of interest for weapon purposes.

The physical and chemical properties of the superheavy elements are also of interest for weapon production, and we consequently ask: What are these properties?

According to theoretical predictions element 110 is eka-platinum, element 111 is eka-gold, element 112 is eka-mercury, element 113 is eka-thallium, and element 114 is eka-lead. Even though the physical and chemical properties are not necessarily very similar to their lower homologues, it is still probable that they are all metals and therefore suited for manufacture of nuclear weapons.

The fourth question to be raised is: Can the superheavy isotopes be produced artificially?

Immediately there seem to be two possible roads to the production of superheavy elements, namely, by the use of accelerators and reactions between charged particles or by the use of successive neutron capture in existing heavy nuclei and connected radioactive decay. The latter approach could be accomplished in nuclear reactors or explosions. Of these two roads, only the first one seems to offer realistic possibilities. The reasons for this conclusion are the following.

High-flux reactors have been very useful in the production of transuranium elements and have produced heavy isotopes up to Fm^{257}. Theoretically it should be possible to go further, but it is extremely unlikely that the island of stability could ever be reached, as the heavy nuclei, produced in the instability gap between the presently known heavy nuclei and the island of stability around ${}_{114}X^{298}$, would most probably undergo spontaneous fission before they could reach the island of stability. It should be remembered that the production of very heavy isotopes is a slow process even in reactors with a very high flux.

Nuclear explosions have also been used to produce transuranium isotopes up to Fm^{257}, and theoretically the possibility exists to go further. Even though the successive neutron captures occur much faster here than in nuclear reactors, the extremely neutron-rich nuclei produced will still have to undergo a series of radioactive decays before they can reach the island of stability. During these decays they may well disappear, owing to spontaneous fission. Even if this approach was feasible, it would result in an extremely low yield of the superheavy elements since only very few nuclei of sufficient mass would be formed. It is worth noting that when a U^{238} target was exposed to neutrons from the MIKE test explosion, the ratio between the number of nuclei produced with $A = 239$ and the number of nuclei produced with $A = 255$ was of the order of 10^{11}.

Thus we can discard the successive neutron capture approach and turn to the accelerators.

One difficulty in the production of superheavy nuclei by accelerators is that since these nuclei are very neutron rich, the particles reacting in the accelerator must also be neutron rich, i.e., be heavy nuclei themselves. Various reaction schemes have been suggested, e.g.,

$$U^{238} + Xe^{136} \rightarrow {}_{114}X^{298} + Ge^{72} + 4n^{1},$$
$$U^{238} + U^{238} \rightarrow {}_{114}X^{298} + Yb^{170} + 8n^{1}.$$

Of these two reactions the second one is believed to have the best prospects.

Unfortunately these reactions cannot be achieved in existing accelerators. Since both reactants are heavy nuclei, the projectile ions must be accelerated to very high energies to overcome the Coulomb barrier, and the ions must be multiply charged to allow the acceleration to be carried out in a machine of reasonable size. The ion energy necessary to accomplish the U^{238}, U^{238} reaction is about 1.5 GeV.

It is also important to realize that such reactions will lead to a whole spectrum of reaction products of which only a very small fraction might possibly be the relevant superheavy isotopes.

Even though these reactions cannot be achieved today, there seems to be little doubt that some time in the future it will be possible to build the heavy-ion accelerators required for these purposes. This could be done either by further development of existing accelerator types or by the use of new concepts. One new concept is the HIPAC machine, in which the acceleration is accomplished by the use of the potential well formed by an electron plasma. This well will both accelerate and contain the ions injected into the machine. Since multiply charged ions are also formed by stripping of the ions in the plasma, the size of the machine should be reasonably small.

According to the original paper on the HIPAC concept, the total reaction rate should be 2×10^8 reactions per second for a typical HIPAC machine and for the U,U reaction. Even if only a small fraction of these reactions results in the formation of superheavy nuclei, that should be quite sufficient for their detection and identification.

The HIPAC concept has not yet been used in an accelerator, but even if it should fail to work, the advanced versions of conventional accelerators or possibly other new concepts would still be available.

Consequently it seems probable that the superheavy elements can be produced artificially by means of accelerators in a not-too-distant future.

It is not enough, however, that these elements can be produced. For our considerations it is necessary that it should be possible to produce them in quantities sufficient for weapon production. As will be shown later, this means that the minimum annual production must be of the order of tens of grams. Therefore our fifth question is: Is such a production possible and if so, at what cost?

One gram of a superheavy element with a mass number of approximately 300 contains 2×10^{21} atoms. If every reaction in the HIPAC machine considered here should result in the production of one usable superheavy nucleus, an overoptimistic assumption, then the production in one machine would amount to only 3×10^{-6} g/yr.

HIPAC machines with much higher reaction rates might conceivably be designed. If on the other hand the reaction yield of usable superheavy isotopes is taken into account, it seems likely that thousands of HIPAC machines would have to be built to get an annual production of superheavy elements in gram quantities. Such an enterprise would demand a very great effort, but might just be possible for the superpowers.

It is unlikely that the advanced HIPAC machine we are considering here could be built at a cost of less than $10 million. Consequently if 1,000 machines are needed for a production of 1 g of superheavy elements (SHE) per year, then owing to interest and depreciation the capital cost would amount to something like $1,000 million per g SHE, a prohibitive price.

Other accelerator types may be used. It is, however, unlikely that advanced versions of existing accelerators will be better than the HIPAC for the acceleration of heavy ions. Entirely new machine concepts may be developed, but it is obviously not possible to evaluate the merits of such new concepts here.

Since the cost question is of great importance, let us try a different approach to a cost estimate. Let us assume that all the energy consumed by the accelerator is used to give the relevant particles the necessary energy for reactions, that all accelerated particles will react, and that each reaction will result in the production of one relevant superheavy nucleus. These assumptions are certainly over-optimistic by several orders of magnitude. Let us further assume that in order to achieve one reaction we have to supply an amount of energy of 1 GeV. As mentioned earlier this is a reasonable assumption.

Based on these assumptions, the energy consumption per gram of superheavy elements produced is found to be 10^5 kWh per g SHE. If the cost of electricity is taken to be 4 mills per kWh, the cost of the superheavy elements is $400 per g SHE.

This result is at least several orders of magnitude too low, not only because of the assumptions made above, but also because it takes into account only the cost of the electric energy consumed by the machine, not the other operational costs and the capital cost. With the machine unspecified it is impossible to say how much higher a realistic total cost should be, but it seems inconceivable that it could be lower than $1 million per g SHE. Probably it is much higher. Consequently the answer to our fifth question is: Production of superheavy elements in relevant quantities may be just possible. However, the cost is likely to be higher than $1 million per g SHE.

Compared with the cost of plutonium, which is around $50 per g weapon-grade plutonium, the cost of the superheavy elements is definitely very high. The cost itself is, however, not sufficient reason to discard any possible military significance of the superheavy elements. It is also necessary to take into account the amount of these elements to go into the production of one SHE-weapon. Hence our next question is: How many grams are needed for one SHE-weapon?

This amount can be calculated provided reasonable estimates of relevant neutron cross sections and atomic densities of the superheavy elements can be made.

The chain reaction in the weapon must be based on fast, unmoderated neutrons in order to achieve an explosion. This means that the neutrons will have energies in the MeV region. Since the relevant cross sections are fairly energy-independent in this region, it does not matter too much if the energy of the neutrons from fission in superheavy nuclei differs somewhat from the neutron energy of fission in, for example, Pu^{239}.

On the basis of extrapolations of the cross sections of known heavy isotopes it is estimated that the superheavy elements have roughly the following cross-section values:

$$\sigma_{total} = 9 \text{ barn}, \qquad \sigma_{f} = 3 \text{ b}, \qquad \sigma_{c} = 0 \text{ b}.$$

As mentioned already, the number of neutrons emitted per fission, ν, is around 10.

The atomic density of the superheavy elements has also been estimated from an extrapolation of the densities of known elements, and for $Z = 110$ to $Z = 114$ it was found to lie in the interval 0.03×10^{24} to 0.07×10^{24} atoms/cc.

By use of these figures a number of very rough critical-mass calculations based on collision probability were performed for bare superheavy-element spheres. The results of these calculations are given in Table 1.

It is seen that for the bare spheres, critical masses in the range of 25 to 500 g are calculated. However, the critical mass is not identical to the amount of fissionable material in a weapon. At the start of the detonation of a nuclear weapon it must contain a strongly supercritical configuration of the fissionable material. On the other hand, a reasonably sized reflector and compression of the fissionable material by the implosion process may significantly reduce the amount needed of this material. The critical mass of a bare Pu^{239} sphere is 16.4 kg, while it is often assumed that the amount of Pu^{239} in a nuclear weapon is around 3 to 4 kg.

If we assume that the critical mass of a bare sphere of superheavy elements is around 80 g and

Table 1 Critical Masses of Bare, Superheavy-Element Spheres.

A. $\nu = 10$, $\sigma_t = 9$b, $\sigma_f = \sigma_a = 3$b, $n = 0.05 \times 10^{24}$ at/cc, $\rho = 25$ g/cc.
Critical Diameter $D_c = 1.8$ cm
Critical Mass $M_c = 80$ g

B. Variation of ν:			
ν	8	10	12
D_c (cm)	2.3	1.8	1.5
M_c (g)	160	80	45
C. Variation of n:			
n (at/cc $\times 10^{24}$)	0.03	0.05	0.07
ρ (g/cc)	15	25	35
D_c (cm)	3.0	1.8	1.3
M_c (g)	220	80	40
D. Variation of σ_f:			
σ_f (b)	1.5	3	4.5
D_c (cm)	3.4	1.8	1.2
M_c (g)	520	80	25

that in a weapon this amount can be reduced by a factor of 5 by reflection and compression, the amoun contained in a weapon is around 15 g SHE. This would mean that the fissionable material in a SHE-weapon is roughly a factor of 200 as small as that of a plutonium weapon, but the price of this material is $15 million in a SHE-weapon (assuming a price of $1 million per g), while in a plutonium weapon it is only around $150,000. The cost of a SHE-weapon is consequently two orders of magnitude as high and probably more.

It is also worth remembering that with a charge of fissionable material of only 15 g, the maximum explosive yield of such a weapon will be much lower than the maximum yield of the plutonium weapons considered. The energy release of a superheavy-nucleus fission has been estimated to be around 30% higher than the energy release in a Pu^{239} fission. However, since the superheavy nuclei are also around 30% heavier, the energy release is roughly the same per mass unit in the two cases.

The complete fission of 1 kg Pu^{239} results in an energy release equivalent to 20 kt TNT. Hence a SHE-weapon containing 15 g of fissionable material has a maximum yield of 300 t TNT. In practice the maximum yield would probably be limited to around 100 t TNT, since not all superheavy nuclei will undergo fission before the chain reaction is stopped.

The results given above are summarized in Table 2, where a plutonium weapon has also been considered for comparison.

It should be mentioned that the weapon weights and diameters of Table 2 are "guesstimates." The weapon diameter is determined not only by the amount and shape of the fissionable material but also by the thickness of the reflector, casing, etc. The weapon weight is the total weight of the weapon.

From Table 2 it is obvious that SHE-weapons are extremely expensive. In comparison it may be mentioned that the USAEC price for peaceful nuclear explosives (including related services) is $350,000 for a 10 kt TNT device and $600,000 for a 20 Mt TNT device. Also, the yield of SHE-weapons is very modest. Their only advantage seems to be their smaller diameter and weight.

The last question is: In view of the above considerations what is the military significance of the superheavy elements?

On a cost-efficiency basis there seems to be no

Table 2 Comparison Between SHE- and Pu-weapons.

	SHE	Pu
Mass of Fissionable Material (g)	~ 15	~ 300
Weight of Weapon (kg)	1–10	10–100
Weapon Diameter (cm)	4–6	20–25
Yield (kt TNT)	~ 0.1	~ 20
Cost of Fissionable Material ($ millions)	$\gtrsim 15$	~ 0.15

reason to produce this type of weapon for ordinary military use, whether as pure fission weapons or as triggers for thermonuclear weapons. The lower weight of the SHE-weapons will permit cheaper means of delivery, but this cost reduction is far less than sufficient to compensate for the increased cost of the weapons.

One might think of very special applications of nuclear weapons where the high price might be justified. One possibility is the use of SHE-devices as triggers in clean thermonuclear weapons. However, these devices may not be sufficiently powerful to ignite the thermonuclear reactions. Also, fairly clean nuclear weapons already exist, and it is highly questionable whether the reduction in the release of radioactivity which could be achieved would justify the high extra cost. Another possibility might be the so-called briefcase bomb, a weapon to be smuggled into enemy territory. A plutonium weapon might also be smuggled, but owing to its larger weight and size it might well be necessary to divide it up into several parts, thus increasing the risk of detection. On the other hand the SHE-weapons, too, have their shortcomings for such use. Since they are likely to be radioactive with half lives not longer than around 10^4 years and possibly significantly shorter, and since the decay products formed are likely ultimately to lead to spontaneous fission, the radiation from a SHE-weapon and the heat generation in the weapon could be a much more serious problem than in a plutonium weapon. These shortcomings would increase the risk of detection and would of course also affect the ordinary operational use of SHE-weapons.

It is impossible to foresee all conceivable military applications of nuclear weapons, now and in the future. It might therefore be maintained that certain military needs might arise which would justify production of SHE-weapons in spite of their high price. It should, however, be remembered that the production of superheavy elements for weapon purposes is a very great technological and economic enterprise, which will only be undertaken if there are major advantages to be gained. And SHE-weapons will by no means add a new dimension to the nuclear arsenals. They will, as I see it, only add somewhat to their versatility.

It may also be said that in this paper a number of assumptions have been made that are too conservative. For example, the amount of superheavy elements needed for a weapon may be lower than estimated here. This may be so, but it is most unlikely that it is orders of magnitude lower. Further, the estimated cost of $1 million per g SHE is, as I see it, very optimistic and should probably be much higher.

Thus in answer to our last question the following can be said: The SHE-weapons will not add new dimensions to the nuclear arsenals, and to start their production would be a very large enterprise. Further, from a cost-efficiency point of view they cannot compete with plutonium weapons. Consequently it is very unlikely that they will have any military significance in the foreseeable future.

Production of Superheavy Elements in Accelerators: Comments by R. Ramana

While we now have no doubt as to the possibility of the existence of superheavy nuclei, there seems to be some doubt as to whether such nuclei can be formed at all by means of accelerators. Some recent calculations of Kapoor et al. to appear in *Physical Review Letters* have shown, from a consideration of the level-density structure available from fission studies, that at fairly low excitation energies (of the order of 10–20 MeV) shell-structure effects are reduced very rapidly with excitation energy. If this is so, since superheavy nuclei can only be formed by means of accelerators with excitation energies of at least this order, the very forces which go towards holding such nuclei together will have disappeared. It therefore seems unlikely that superheavy nuclei can ever be made by means of accelerators.

Nuclear Physics for Peaceful Uses: Comments by O. Kozinets

The achievements of nuclear physics include not only the peaceful uses of nuclear forces, as in chain reactions, but also the gloomy applications to new and powerful military systems. The energy released by an atomic explosion is 10^6 times greater than that released by TNT. However, no system of units exists which can measure the depth of human grief or the bitterness of mothers' tears which have resulted from these weapons. The creators of the atomic bomb were not able to imagine all the consequences, present and future, of their work. The gloomy echoes of the Hiroshima and Nagasaki explosions are still heard all over the globe, even today. Nothing can stop or damp them. Instead of being used for the common good of all people all over the world, atomic energy has been used to create fear and tension for all humanity. Many well-known physicists have looked back with horror at their roles in the development of atomic weapons.

It is with deep respect that we pay tribute to the memory of the late Niels Bohr. His wisdom and humanity can serve as a good example for all of us. First in England and later in the United States Bohr had been connected with the atomic project since 1940. He took an active part in the taming of nuclear chain reactions up to June 1945. At a time before an atomic bomb had actually been constructed, Bohr sent a memorandum to Roosevelt. In August 1944 he wrote that the unity among the Allies created in their fight against fascism would dissolve after the war, and he insisted on some actions which would guarantee world security, such as the complete banning of nuclear weapons through international control. After the disastrous war Bohr continually thought about the future of atomic energy and worked for international cooperation in science for the benefit of humanity.

The discovery of nuclear fusion, like many other

discoveries, was made as a result of the irresistible aspiration of human science toward self-expression and new knowledge. However, in spite of these noble motivations, the first practical realization was in a military application, in the form of a hydrogen (thermonuclear) bomb.

A controlled fusion reaction is presently still in the research stage. However, judging by the recently announced plans by Seaborg, our American colleagues have made a decision to tackle this problem seriously in the manner of the Tokamak construction. Let us wish them all success.

The main goal of our symposium is to discuss some aspects of the development of technology and possible new discoveries in the future. It is not easy to make a reliable forecast. Human forecasting usually has been very shortsighted, particularly in science. There are many historical examples of incorrect estimates made by great scientists. Some of them changed their views from scepticism to enthusiasm, then worked very hard to solve the problem. This has been the way of many revolutions in science.

At any rate let us try within the limits of our intuition and imagination to develop schemes which might warn the public and perhaps avoid further madness in the world.

It seems that the key issue in the armaments race is fear, hidden deep within men. Scientifically it has been expressed in the so-called worst-case analysis (WCA) as it is applied to strategic arms. If we begin with a WCA there will immediately follow additional expenses to cover the fear-gap. But each step gives no one substantial superiority in the strategic arms balance because of the assured destruction capability, regardless of who strikes first. Who then gains? Those people connected with the weapons industry-military complex gain the greatest benefits. There are some specialists who use WCA for obtaining military contracts in order to support their laboratories and for huge new installations. Everybody understands that it is the smallest part of the military budget that is spent on such kinds of scientific activity but that it is this scientific activity which leads to the huge burden of the useless and dangerous arms race.

Any weapon is closely connected with different kinds of scientific achievement, and all practical discoveries in science can be used for military applications. Let us here limit the problem by taking into account only those scientific directions which can change the existing strategic balance between two great powers.

Contemporary international affairs are characterized by a state of dynamical balance between the great powers. The number and quality of armaments has been steadily increasing on both sides and may be described by complicated functions of many variables, one of them being scientific potential. The majority of our symposium members have accepted the reality of the fact that each side has enough strength to retaliate in the worst case and that a first strike would be tremendously dangerous for both sides. In such a situation offensive hostilities with the use of atomic weapons can only result in self-destruction.

Returning to our subject of discussion, we must analyze any possible scientific or technical breakthrough that could radically change the military strategic balance.

To simplify the problem we may say that any weapon is directed to destroy its target by releasing a sufficient amount of energy in a predetermined place at a certain time. The weapon consists of a warhead, a booster, and a system of guidance or navigation. Strictly in terms of nuclear physics, the amount of energy provided by the use of a nuclear or thermonuclear device is more than enough for any target. The energy yield from a nuclear warhead cannot be greater than the theoretical value calculated on the basis of the mass defect. It is possible to expect some improvement in the efficiency of atomic warheads in the range of less than an order of magnitude. Nevertheless, these changes cannot influence the strategic balance.

Taking possible combinations of elements which occur in different nuclear reactions should not give any substantial increase. The final yield will be of the same order of magnitude.

There are also some speculations about the use of annihilation processes for military purposes. However, there are no foreseeable developments which can give positive results in the collection and confinement of a reasonable amount of antimatter.

Nuclear physics, then, from an energetic point of view, cannot bring any new surprises.

One could consider the possibility of creating some kind of radiation weapon. Not being a specialist on lasers I will only touch on this subject. In principle, however, it is theoretically possible to have a γ-laser. The only question up till now is how and which combination of elements could be used for it. It is quite possible that a deep penetration into the theory of the structure of nuclei and future developments in γ-spectroscopy and the Mössbauer effect will bring some positive results in the way of constructing a γ-laser.

In conclusion, we must avoid a fear of each other and propogate an understanding of the nature and causes of the arms race. We must also cooperate with one another in different branches of human activity. As a good example of such cooperation we could work on high energy accelerators and elementary particle physics.

Summary of Discussion

Requirements of Long Half Life. The reason that a long spontaneous-fission half life is necessary for the material used in a fission bomb is that in the implosion process, if the neutron background is too high, the chain reaction will be pretriggered and a fizzle will result.

Difficulties in Working with Superheavy Elements. The characteristics required of a material to make it useful for weapons are quite stringent. Plutonium, with an alpha-decay half life of 2×10^4 years, when wrapped in high explosives for an implosion device is about 20° to 30° hotter than its surroundings. With the shorter half lives contemplated for the super-heavy elements, this temperature difference would likely be too great to make the material usable for weapons unless a cooling system were employed. Similarly, the radiation levels would make the material very difficult to handle.

Superheavy Elements Used to Support Normal Fission. The superheavy elements could possibly be used, because of their high neutron yield, to support the normal fission process. For this purpose only very small quantities need to be used. Still, the cost of producing superheavy elements is so great that the optimum cost-effectiveness ratio may well dictate the use of 100%, or nearly 100%, conventional fissionable material.

Manufacture in a HIPAC. Experiments at the Avco-Everett Laboratory on the design of a HIPAC machine had progressed far enough, before funds ran out, to demonstrate that the necessary high electron densities and containment times were probably achievable and therefore that in principle the HIPAC would work. The estimated cost of $10 million may not be far wrong for the first working HIPAC, but it is a sufficiently simple machine so that, with a cheap source of power, the cost of construction could be brought down considerably. It is not likely, however, that the price of the superheavy elements will come down close to the stated lower limit of $400 per gram, especially because the yield factor will probably be low. But if this price could be brought down by a factor of 50 or so from the estimated $1 million per gram, then except for the remaining handling problems, the possibility of mixing the superheavy elements with conventional fissionable material may be realized. On the other hand the $1 million estimate was arrived at by taking a 10% yield factor, an efficiency of about 1% to 10%, and an overall unaccounted-for factor of ten for other costs, including possible isotope separation and capital costs. Therefore, $1 million per gram could well be an underestimate.

Location of the Island of Stability. There have been a number of different predictions of the exact location of the island of stability (see, for example, A. K. Kerman and W. H. Bassichis, "Self Consistent Calculations of Shell Effects Including the Proposed Island of Stability," submitted to *Physical Review*, March 1970, where they predict that it falls near $Z = 120$). Dr. Olgaard indicated that the results of his paper would not be sensitive to this uncertainty of the location of the island of stability.

Nuclear Weapons Technology

J. Carson Mark

The basic processes involved in nuclear weapons are those of fission and the burning of the thermonuclear fuels, D, D-T mixture, and LiD. The energy from fissile material fully consumed is about 1.7×10^7 times as large as that from the same mass of high explosive. The energy from fuel burning of the various thermonuclear fuels is only 3 or 4 times larger than that of the same mass of fissile material. The thermonuclear fuels have the additional difference of providing up to 6 times more free neutrons than fission for the same energy. Apart from this, the principle differences are that the thermonuclear fuels—some of them, at any rate—are relatively abundant and inexpensive; and they may be assembled in arbitrarily large amounts, without being subject to such limitations as the need for a critical mass (see Table 1).

These fuels are all there are in sight, and there seems to be no prospect of finding others with importantly different capabilities. They have already been used in various combinations and with good effectiveness and there is no reason to suppose that new patterns and combinations can be formed which will change in an important way the general range and type of nuclear weapons or nuclear-weapon effects beyond those already available. Improvements, adaptations, modifications—yes; fundamental changes—no.

J. Carson Mark is Director of the Theoretical Division, Los Alamos Scientific Laboratory.

Table 1 Comparison of Fission and Fusion Reaction Properties.

		MeV/Mass Unit	MeV/Neutron
Fission Reactions:			
	$n + Pu^{239} \rightarrow 180$ MeV $+ 3n$	$180/240 \approx 3/4$	~ 100
Thermonuclear Reactions:			
a) D-T	$D + T \rightarrow He^4 + n + 17.6$ MeV	$17.6/5 \approx 3.5$	17.6
b) Li^6D	$n + Li^6 \rightarrow He^4 + T + 5$ MeV $D + T \rightarrow He^4 + n + 17.6$ MeV	$22.6/8 \approx 2.8$	?
c) D	$D + D \rightarrow n + He^3 + 3$ MeV $D + D \rightarrow p + T + 3$ MeV $D + T \rightarrow He^4 + n + 17.6$ MeV	$24/10 \approx 2.4$	~ 12
d) As above plus	$n + He^3 \rightarrow p + T$ $D + T \rightarrow He^4 + n + 17.6$ MeV	$42/12 \approx 3.5$	~ 20

Energy/Kilogram (Thermonuclear) ~ 3 or 4 times Energy/Kilogram (Fission)
Neutrons/Energy (Thermonuclear) ~ 1 to 6 times Neutrons/Energy (Fission)

It is useful to recall that the first fission weapons—about 20 kilotons in about 10,000 pounds—gave an increase in yield per pound of weapon weight by a factor of about 4×10^3 above that previously available from high-explosion weapons. Because of the possibility of using arbitrarily large amounts of fuel, the thermonuclear weapons as first realized in the United States, and as realized by the USSR in their tests in 1961 when they produced the largest explosion ever conducted (60 megatons), showed an increase in the yield of a factor of 10^3 larger than the early fission weapons. As mentioned earlier, however, this factor was not in the yield per pound of weapon weight, but rather in the weight of fuel it was possible to assemble. These very large devices are not at all suitable for carrying on rocket carriers and so far as one may assume are not a part of anyone's current weapon systems.

The main thrust of weapons work in the past 10 or 15 years has been to obtain effective weapon packages at weights suitable for rocket vehicles—as single warheads in the first instance, and as units for multiple warheads in more recent instances. Efforts along these lines have been quite successful, to the extent that improvements in yield per pound of something like a factor of 10^2 over the first fission bombs have been realized. This brings the yield per pound up to a few times 10^5 larger than high explosion, which is to say close to within an order of magnitude of the ideal, hundred percent advantage factor of 1.7×10^7 (for fissile material without any assembly hardware) or 3 or 4 times larger for bare, unpackaged thermonuclear fuel. In this respect therefore, as Dr. Kozinets has pointed out, larger or really important improvements are not available—and need not be feared.

The possibility that superheavy elements (as discussed by Dr. Ølgaard) may exist and may have smaller critical mass and smaller weapon weight cannot be absolutely ruled out; but it is reasonable to suppose that even if these materials have properties which allow their use in practice (which is questionable) and costs which permit them to be considered at all (even less likely), they would not make an essentially important change in the existing situation. Similar comments apply to the possibility discussed by Dr. Brunelli that it may be possible to achieve pure thermonuclear explosions by igniting dense plasmas of D-T by means of laser energy.

In spite of these apparent general facts, much work of practical and theoretical interest remains to be done on weapon design and development. There is much yet to be learned about particular effects (x rays, neutrons, ionization, and blackout) at high altitude. As long as new carriers or systems may be considered, new adaptations of weapon packages will be required. From the particular point of view of new and improved weapon systems, many useful and interesting improvements in weapons are certainly possible. But none of these constitute significant changes in the nature and scope of the problems already presented to human society by nuclear weapons.

What has been said above applies specifically to countries such as the United States and the USSR and possibly a few others who have had nuclear weapons in hand for some time and who have worked hard to adapt such devices for handling on the relatively newer types of carriers, and for multiple warheads on such carriers. The really large qualitative change in the nuclear-weapons picture seems most likely to come from the possibility that many nations not presently possessing nuclear weapons of any type may come into the position of obtaining some such weapons—even if only of unsophisticated types. In this connection I should like to mention a comment of Dr. Prawitz of the National Research Institute of Defense in Stockholm to the effect that a colleague of his has become persuaded that he could produce a nuclear explosion from essentially any grade of reactor-produced plutonium that might be available. I am not familiar with the details of the calculations of Dr. Prawitz's colleague, and I can only assume that by "nuclear explosion" he means what I would mean, which is an explosion of at least 3 orders of magnitude more energy per pound than would be available from high explosives. From my own considerations of this problem, I have no reason to question such a conclusion, and I would like to warn people concerned with such problems that the old notion that reactor-grade plutonium is incapable of producing nuclear explosions—or that plutonium could easily be rendered harmless by the addition of modest amounts of the isotope Pu^{240}, or "denatured," as the phrase used to go—that these notions have been dangerously exaggerated. This observation is, of course, of no direct practical interest to the United States or the USSR, who have adequate supplies of weapon-grade plutonium, and have proved designs for weapons much better than could easily be made with plutonium from power reactors. To someone having no nuclear weapons at all, or no source of high-grade materials, however, the prospect of obtaining weapons—even of an "inferior" or "primitive" type—could present quite a different aspect.

Summary of Discussion

Radar Blackout. There seems to be no reason why an offense could not cause a very-high-altitude burst which would result in a large and long-lived plasma cloud. Such a cloud could, in all likelihood, effectively hide the immediately following missiles from the long-range radar until they came out the other side of the cloud. This concealment could reduce the time available for a long-range interceptor to fly out to its target and thereby render the long-range defense ineffective. However, if it were not for the additional problem of lightweight penetration aids, enough work might be done on this problem to overcome it. In the case of the terminal defense, the seriousness of a low, atmospheric burst depends very much on the exact design of the defense. In general, ABM designers think they can engineer around this problem.

Complete Test Ban Treaty. Dr. Mark was asked to speculate about the effect of a universal test ban treaty on the weapons programs of the major powers and on the possibility of proliferation. He suggested that in his opinion the configuration of weapon systems in the United States would soon be frozen, employing only those devices which had actually been tested. There are two options available in making a nuclear device for a new missile. A new device can be built to fit the desired missile specifications. This is often possible because the lead times on nuclear devices are short compared with the lead times of missiles. However, the missile can be built to accept some device which is already in the inventory. In the event of a comprehensive test ban the latter course would probably be followed. On the other hand, work could still continue in certain theoretical areas like the effects of blackout or the electromagnetic pulse phenomena. There would probably also be pressures from time to time to abrogate the treaty in order to test new ideas. The United Kingdom has conducted few tests lately and during this period has not deployed any new systems. It is also having difficulty keeping an active group working in the field.

With respect to the proliferation question, there are some kinds of crude weapons, like the gun-assembly weapons, which a country could be sure would work without prior testing. Plutonium, however, could not be used successfully in such a gun assembly. Since the nonproliferation treaty imposes a test ban on nonnuclear countries, this is presumably the situation that exists today for those nonnuclear countries which subscribe to it.

High Energy-Density Plasmas and Pure Fusion Triggers

B. Brunelli

The interest of the author in this subject is due to its connection with the research on controlled thermonuclear fusion.

This report, therefore, is biased by the reasoning involved in discussing the necessity of harnessing the released energy, taking account of economical considerations. Secondly, it will reflect a complete ignorance of classified matters and will refer to results obtained in open laboratories.

Introduction

It is well known that to get thermonuclear energy of practical interest, it is necessary (but not sufficient) to heat heavy hydrogen to a temperature T greater than 10^8 degrees K (ignition temperature) for a time of confinement

$$\tau > \frac{1}{n} 10^{14} \text{ sec}$$

for the most convenient mixture of deuterium and tritium, with particle density n (cm^{-3}). This condition follows from (using self-explanatory notations)

$$\frac{E_{out}}{E_{in}} = \frac{3nkT + \dot{W}_F \tau}{3nkT} = A > 1,$$

where $\dot{W}_F = \frac{1}{4} n^2 \langle \sigma v \rangle Q$ and $A = g + (1/\eta)$, g being the useful gain of energy and η being the

Bruno Brunelli is Director of the Laboratorio de Gassi Ionizatti at Frascati, Italy.

product of the efficiencies in transferring the (circulating) energy from the source to the plasma and vice versa. One finds

$$\tau = \frac{(A - 1)\ 3kT}{\frac{1}{4}\, n\, \langle \sigma v \rangle\, Q}\ .$$

In trying to satisfy this condition (of Lawson), one has evidently two main final possible goals: a quasi-stationary fusion reactor (long τ, low plasma density); or a pulsed reactor (short τ, high density), which requires *high energy-density* plasmas (e.g., $nkT > \text{kJ/cm}^3$).

A possible impact on the arms race of the hypothetical quasi-stationary fusion reactor could be the use of its huge flux of neutrons to breed plutonium for nuclear weapons in a uranium blanket; but this should imply an evident change in the physiognomy of the peaceful reactor.

The control of the energy from a hypothetical pulsed fusion reactor consists in succeeding to release an amount of energy that can be contained in a vessel or in a cavity; if the aim is electricity production, the cost of the damaged parts has to be so low as to assure competitiveness with the conventional methods of electricity production. This economic condition is rather severe; it is offset by the advantage that this hypothetical pulsed reactor does not suffer severely from plasma instabilities.

If these restrictive economic conditions are abandoned, the violent release of pure thermonuclear energy from the hypothetical uneconomical pulsed reactor could have special applications.

To contribute to the evaluation of how hypothetical is such a reactor, I will describe in the following sections the status of the researches on high energy-density plasmas. For references, see the exhaustive review article of J. G. Linhart (Ref. 1).

The complexity of the devices for the production of such a type of plasma renders the devices, in case of success, rather difficult to transport and to handle. In my opinion, in order to evaluate realistically whether some special applications can have military use, it is necessary to evaluate how much the complexity can be reduced.

Laser-produced Plasmas from a Solid Target

Among the approximately ten open laboratories interested in very hot and dense plasmas, some have succeeded in producing several thousand neutrons by focusing a powerful beam of laser light onto a solid pellet of deuterium or LiD. The characteristics of the neodymium lasers used for getting those few thermonuclear neutrons are quite impressive; the energy of the beam of light is of the order of 10 J and in one case 200 J; the duration of the pulse ranges from tens of picoseconds (10^{-12} sec) to tens of nanoseconds (10^{-9} sec); in this range one can find a duration sufficiently shorter than the time for an appreciable expansion of the target and nevertheless sufficiently long for a significant equipartition of energy between electrons and ions.

Not only lasers, but probably also electron beams and hypervelocity projectiles (grams at 10^7 cm/sec) can deposit rapidly large quantities of energy into a fusionable target. *Mode-locked lasers* provide, at present, the highest fluxes of energy (announced at Lebedev Institute, Moscow; at the University of Rochester; and at UCRL, Livermore: lasers of 1 KJ for nsec or 100 J for psec on areas smaller than 1 mm^2).

Higher values of the total beam energy distinguish the *electron beams*: the best performance quoted to date (*Physics Today*, June 1969, p. 59) is 10^6 amperes, 10^7 volts for 10^{-7} sec. *Hypervelocity projectiles* (produced in macroparticle accelerators) are just under consideration. For references see the review article of J. G. Linhart (Ref. 1).

Considering the laser as the source that at present gives the best overall performance in releasing its energy to a solid fusionable target, one can say that this energy is by many orders of magnitude lower than that needed for triggering sufficient reactions for a positive energy balance. In quoting the minimum input energies E_{in}, I will distinguish two cases:

(*a*) E_{in} *heats directly part of the fuel mass.* A powerful laser beam is focused on a solid 50/50 D-T mixture and creates a hot spot in the mass. The energy E_{in} given to the spot can provide in the surrounding mass a detonation wave fed by the energy deposited mainly by the α particles (of 3.5 MeV) produced in the fusion reactions. The minimum energy E_{in} required for triggering the detonation has been calculated with different models and, correspondingly, with different results (see, for instance, Refs. 1–3); anyhow, E_{in} ranges from a few hundred to few thousand MJ (10^6 J). According to Caruso (Ref. 3) the trigger energy could be reduced by a factor of about 500 if it were possible to precompress the pellet by a factor of 5, to use a tamper and a magnetic field of 5×10^5 gauss (in order to reduce the thermal conductivity); that is, the trigger energy with these rather elaborate conditions would be of the order of 10^6 J. The energies released by the best lasers presently in use for this purpose are at least 3 orders of magnitude lower than 10^6 J.

(*b*) *E_{in} heats the whole mass of the fuel.* In this case the target is a D-T solid pellet which is brought to the ignition temperature by the laser beam focused on it. The energy E_{in}, required for the release of an energy $E_{out} = AE_{in}$, is proportional to the volume (r^3) of the pellet and hence to $(A - 1)^3$ because r is proportional to τ, which is proportional to $(A - 1)$. Taking this into account, it is sufficient to report evaluations of E_{in} made for a particular value of A, say $A = 10$. In the case of *free expansion,*

$$E_{in} = 10^9 \text{ J},$$

delivered in 10^{-10}–10^{-9} sec; $r \sim 1$ cm; explosion containable in a cavity of 9 m in radius. (Data derived from Refs. 2, 3.)

In the case of *tamping,*

$$E_{in} = 7 \times 10^7 \text{ J},$$

with $r \sim 0.4$ cm; energy easily containable. (Data derived from Refs. 2, 3.)

In the case of *tamping + strong magnetic field,*

$$E_{in} = 14 \times 10^6 \text{ J},$$

with $B \sim 3 \times 10^5$ gauss in order to reduce thermal conductivity in the radial direction of a cylindrical geometry. (Data derived from Ref. 3.)

The figures quoted in case (*b*) suggest the following considerations: (1) The E_{in} required in the most favorable case (14×10^6 J) is very far beyond foreseeable progress in laser technology. (2) That the quantity $A = g + (1/\eta) = 10$, does not mean necessarily an overall positive energy balance ($g > 0$); for this an efficiency $\eta > 1/10$ in the laser operation should be necessary. The lasers in use at present are far from this requirement.

Plasma Focusing

A magnetic field, generated by the current flowing through a plasma, can compress it to very high densities and temperatures. In devices developed in several countries (following a Russian scheme), the plasma is concentrated for about 0.1 μsec in a volume of 10^{-2}–10^{-1} cm^3 at a density of the order of 10^{19} cm^{-3} and at a temperature of the order of 1 KeV. With devices fed by condenser banks of about 200 KJ, bursts of 2×10^{11} neutrons per pulse with deuterium (10^{13} with D-T mixture) have been obtained; probably the neutron yield can be increased by a factor of 100 just by increasing the input energy to around 10^6 J.

The phenomena operating in a plasma focus are not yet clear. There are doubts that the present version of the plasma-focusing system could become an economic pulsed fusion reactor, though it is certainly a cheap pulsed source of fast neutrons.

Conclusion

In conclusion, we have to pay more attention to laser-produced plasmas and in my opinion the evaluation of the feasibility of a trigger device depends essentially on the technological development of the energy sources: lasers and electron beams. The evaluation of the manageability of such devices depends on the efficiences of the components involved in the conversion of the different forms of energy required.

References

1. J. G. Linhart, "Very high density plasmas for thermonuclear fusion," to be published in the near future in *Nuclear Fusion*. The review article contains an exhaustive list of references.

2. M. J. Eden and P. A. H. Sanders, "Triggering requirements for pulsed fusion reactors," to be published in the near future in *Nuclear Fusion*.

3. A. Caruso, "Alcune possibilità di applicazione dei laser di potenza in ricerche di fisica," Laboratorio di Gassi Ionizatti Internal Report (1970).

The Use of Intense Relativistic Electron Beams: Comments by R. Z. Sagdeev

In principle, a powerful relativistic electron beam (*Phys. Rev.*, vol. 174 (1968), p. 212) may be used instead of a laser beam for the scheme considered by Dr. Brunelli. Interacting with a solid target, this beam could heat it up to thermonuclear temperatures. The stopping path of the beam should be sufficiently short (in practice less than 1 cm) in order to allow the electrons to deposit their energy into a small volume of material. It may be easily shown that the ranges of single electrons are extremely large, even in condensed matter, since the collision cross section drops with energy as E^{-2}. An acceptable stopping path may be obtained if the interaction between the electron beam and the target material (the latter may be considered as a plasma of very high density) is of a collective nature. This kind of interaction is realized when the beam intensity is so high that plasma oscillations are excited due to the so-called beam instability. In this case the material heating proceeds according to the following scheme: energy of the electrons in the beam → energy of the material plasma oscillations → thermal energy. In general, the process is rather complicated and various authors give essentially different estimates of the ignition energy for a thermonuclear reaction, E_{ign}. F. Winterberg (*Phys. Rev.*, vol. 174 (1968), p. 212) estimates E_{ign} to be of the order of several MJ. B. Breisman and D. Rjutov (*Sov. Phys.—JETP*, vol. 11 (1970), p. 606) made a more detailed analysis of plasma oscillation excitation by an electron beam in the expanding material of the target. They found that E_{ign} is greater than 100 MJ.

Thus E_{ign} for an electron beam is approximately of the same order of magnitude as for a laser beam. At an electron energy of 10 MeV and pulse time of 100 nsec, the beam current should be equal to 10^8 amperes. Existing electron beams (Sandia Laboratories Research Report no. SS–RR–69–421 (1969)) have currents approximately 10^{-3} times less.

Magnetohydrodynamics: Comments by V. Seychev

It is well known that the majority of achievements in military technology are by-products of civil technology. In this connection, it is important for this symposium to analyze not only those achievements of science and technology which already have military applications, but also those new developments which as yet have no military applications. Among these new developments is the magnetohydrodynamic (MHD) method of direct conversion of energy.

The realizing of the MHD method of energy conversion is very important for power engineering because an MHD power plant has a higher efficiency (55%–60%) than a conventional steam power plant (maximum 40%–42%, but ordinarily about 30%). This results in fuel economy and decreasing thermal pollution. Other important advantages of MHD generators are their compactness (liquid-metal MHD generators) and the absence of rotary parts. For these reasons, the MHD generator may be used in the future as a compact source of electrical energy in transportation systems.

These special advantages make the MHD generator attractive for military applications. There have been some publications about military applications of MHD generators both for tactical aims and for strategic aims (for jamming radars and other things).

To make difficult the military use of this (and other) new kinds of technology, it would be very useful to have international cooperation, coordinated research, and exchange of information about investigations in this field. A good example of this is the work of the International Liaison Group on MHD of the International Atomic Energy Agency (IAEA).

Summary of Discussion

Very High Power Lasers. It is not anticipated that a neodymium laser providing energy outputs of the order of 10^6 joules will be built in the future. With components now available, just by increasing the size, 10^5 joules could be reached, but to go higher would be very difficult. The possibility could not be ruled out, however, that chemical lasers could achieve higher energy output. Means of obtaining higher levels of laser performance are under active investigation at many places (see, for example, UCRL Reports 72433 and 72434, and *Physics Today*, July 1970, p. 55), so that it is somewhat rash to assume that large increases in power level may not come rapidly into sight.

High-velocity Macroscopic Particles. The possibility of accelerating macroscopic particles to very high velocities in order to act as a fusion trigger is presently in the area of complete speculation.

Lasers and Electron Beams as Weapons. Lasers or electron beams of high enough power to ignite a fusion reaction could conceivably be used as weapons in their own right. The limiting factors here are the cost and particularly the large size and lack of transportability of these machines.

Reconnaissance and Surveillance as Essential Elements of Peace

E. Fubini

Reconnaissance and surveillance are often considered as tools of war—I think of them as essential elements of peace. "Good fences make good neighbors"— good reconnaissance is an essential element in cooling off the arms race.

Without reconnaissance, all the hawks in both countries will assume or claim that the other side is ten feet tall. The statement will always be made: "they" *could* do "this," therefore, "since we don't know that they are not doing it, we must act as if they did do it. We should, therefore add to our offensive forces. . . ."

This sets the basis for escalation and the continuation of the arms race. It is recognized that reconnaissance reduces the security of the other side if we define security in a narrow sense. But security is often used not to prevent the other parties from finding out about what one has but to hide what one does not have. Edward Teller has recommended that all classifications be eliminated, because the barriers to free propagation of knowledge do more harm to the United States than the advantages that thereby accrue to it. I don't agree with Teller but I have a firm conviction that the more we know about each other the more likely we are to stay at peace.

I am happy to see that progress in technology has made a much wider range of surveillance and reconnaissance procedures acceptable, and I would

Eugene Fubini was formerly Deputy Director, Defense Research and Engineering.

like to spend some time to list present progress, and then to conclude with a section that tries to prove that, without cooperation between countries, even this progress is not sufficient.

Reconnaissance techniques have progressed and we can visualize further progress. High-class pictures from airplanes are possible now; we can record, develop, and retransmit pictures. The lunar orbiter reproduced pictures on film, developed them in space, scanned them with a laser scanner, and sent the images down to earth. Communication satellites are used regularly to transmit pictures from Vietnam.

We can transmit wide bandwidths of many tens of megabits. We can record information at the same rate. The time is not very far off when we will have digital recordings at more than 10^8 bits per inch.

Navigation of our reconnaissance airplanes has improved to the point that high absolute accuracies are available and relative accuracies of 100 ft are possible.

We have sensors that listen to sounds, detect the footsteps of people walking, smell the presence of human beings, and radars that detect trucks; we have MTIs (moving target indicators) on aircraft, we have coherent and noncoherent side-looking radars. We have infrared scanners, image converters, and infrared superheterodynes. We can use all of these sensors in remote-control and remote-reporting systems that we call the "instrumented battlefield."

In the field of nonimaging sensors we have receivers for frequencies to 100 GHz, infrared superhets, noise figures that we expected years ago to be available only at much lower frequencies. We have solid-state microwave radars and receivers. We have learned to analyze complex signals in real time; we have computers that use microwatts of power per bit, solid state memories that cost 10 cents per bit and are projected to cost 1 cent per bit in a few years. Computers with mean time between failures of a year are within reach. We can detect reentry vehicles with our radars; we can measure wake characteristics; we can take ballistic pictures of the reentry vehicle. We have over-the-horizon radars that can see at thousands of miles.

These sensors can be used in a different way: I am convinced that the bad experience of test ban discussion has hidden from us the great usefulness of "transparent black boxes" which would be located anywhere in U. S. or Soviet territory and equipped with a sensor of some kind. The location could easily be checked, the size would be small, the sensors would limit strictly the scope of the information, with both parties fully knowledgeable of the details of the box. A typical transparent box would consist of a camera capable of taking consecutive pictures (say every 30 seconds) of a missile silo to prove that new missiles are not being substituted for old ones.

Partners that are to some extent opponents (politically, militarily, or in other ways) may derive mutual benefit from establishing some sort of a reconnaissance, control, or safeguard system. In order to make such a system work in the framework of mutual partial distrust, it is necessary to have a system that is objective and formalized. It must be objective to be acceptable to both, the inspector and the inspected, and it must be formalized in order to cut the inherent open-endedness of the inspection process: when are you satisfied? Objectivity and formalization may be achieved in various ways. One fairly pronounced way of achieving this is the black-box approach. It is highly formalized because otherwise it would be impossible to design it. One special feature of this point of formalization is its limitation. By the design of the instrument you convincingly install a limitation: certain measurements are possible, others are not. Therefore the receiving partner knows what he is facing. In the situation of partial distrust and therefore limited trust and cooperation, this limitation, which is inherent in the black-box concept, may well be the overriding feature of the concept.

Results of a reconnaissance, control, or safeguard system should not be considered to solve a certain problem completely, as they do not result in a yes/no situation; at least that is true for many cases. For instance, a system for safeguarding nuclear material in the field of peaceful applications of nuclear energy cannot entirely exclude the diversion of very minor amounts of nuclear materials over a very extended period of time; it only sharply reduces the probability of not having detected such diversion. Dealing with probabilities instead of certainties in the full sense again introduces some sort of a limitation. The operational value of reconnaissance, control, or safeguards is principally to be sought in an area that can be called soft. Therefore it is finally not an argument against the black-box concept that it gives only limited information at a limited level of confidence.

It is very important that we realize that despite all the progress in technology, reconnaissance is not good enough to insure the full verification of SALT agreements with purely national means, but that the use of some of the transparent boxes mentioned previously could supply an answer. This would, of course, not be different than the set of buoys proposed by Dr. Mikhaltsev. The only difference is that the buoys in this case may be on land.

Satellites are not as good. There are places in the world where the sky is overcast all the time. Our Scandinavian friends here today can tell me whether it is true that in Bergen, for instance, the sky is overcast 364 days of the year. At high latitudes, the polar night covers the countryside during the winter and ice fog covers the country often during the summer. Satellites have limitations: their size increases roughly as the cube of the desired resolution and, for a given resolution as the square of λ. Thus, for a given performance, doubling the resolution and shifting the spectral band to the near infrared would multiply the size of the satellite by a factor of about 30 and increase the cost correspondingly; the cost of analyzing the data would increase by a factor of 4. This shows why there are practical limits to satellite design. We must also remember that satellites don't look inside houses, nor do they look in the control boxes, nor do they determine intentions.

For instance, I see no way to verify the existence

of a MIRV deployment from a satellite, nor do I see an easy way to answer the problem of the "residual ABM" capability. This is the problem of enforcing an ABM ban while permitting the existence of SAM. Every SAM has some ABM capability because a reentry vehicle with, say, a β of a few hundreds has a terminal velocity identical to those of aircraft and, therefore, a SAM has the *footprint*, although small, of an independent ABM. In addition, most SAMs can be used in the point-in-time–point-in-space method to provide a form of area defense. For these reasons, the type of reconnaissance that existing technologies can provide can do only poorly in verifying the full observance of MIRV and ABM agreements.

As I said at the beginning, I see these verifications as an essential tool of peace because I, and most of my friends in the United States, want to reach arms-limitation agreements and see MIRV and ABM limitations as important portions of these agreements. I see cooperation between countries in permitting the use of cooperative verification by means other than satellites as a possible solution to these problems—the same type of cooperation that led the Soviet ships to lift the tarpaulins off the crates that took the Cuban missiles away from Cuba.

Without going into details, I would like each country to think of methods of verification that would reassure the other party without being either intrusive or indiscreet. I believe that we have in the sensor technology that I described, and in the possibility of verifying the position of a device and interrogating it, the seed of an approach that has been proposed before but never fully thought through. I urge that these ways be explored.

To conclude, reconnaissance is a tool of peace. We have a tremendous range of new technologies available to do both reconnaissance and surveillance, but even our best technology is still limited in its ability to answer some important questions. I believe that further progress of these technologies in a cooperative mode can powerfully contribute to peace.

Summary of Discussion

The ABM Capability of SAM. Every SAM has a "residual ABM capability" although its footprint may be very small. The "footprint" of a SAM missile is that area at which an incoming missile would have to be aimed for the SAM to destroy it. The footprint, therefore, is the area protected. Even the U.S. air-defense missile, Hercules, has a nonzero but small potential ABM capability. Operationally, however, this capability is not realized because the missile is not tied into a system which will allow it to be so used. We must distinguish between a potential capability of a system and the purpose for which it is actually intended at present. In order to determine the intent of a SAM installation, information about the command and control network may be required.

The Use of Black-box Surveillance Devices to Monitor Treaties. There were many discussions in the United States during the test ban negotiations on the possible use of "black boxes" to monitor a total test ban. Unfortunately, as the demands of those who sought complete and absolute assurance of compliance to the treaty were met, these small monitoring boxes grew into unwieldy monstrosities. It is widely believed that in any specific real situation, the same thing would happen again. However, it may be that this experience of the test ban days is now blinding us to the possibility of the feasible employment of black boxes as a means of verification. If the black box were recognized from the beginning to have only a limited function, then it need not grow into a monstrosity. An example of such a limited use for a black box would be the monitoring of a large, sophisticated commercial air-traffic control radar in order to guarantee that it does not have an ABM capability. The properties of limitedness and predetermination which are built into a black box are not possible with air or satellite surveillance, and especially not with human reconnaissance.

It is true that one side could jam such a device and that no one could prevent this, but the box could be so constructed that the other side would know that the box was being jammed. That information alone is sufficient to indicate that something extraordinary is being done.

There is a sense in which the introduction of the concept of black boxes as a means of verifying a treaty would tend to inhibit the negotiations for the treaty. In a strongly bipolar world, where the overall balance of power and deterrence is fairly stable, only relatively gross violations are of interest. This is true both because major political decisions are required to abrogate an arms control treaty and because the international situation is just not sensitive to minor violations. To embark on the road of using black boxes as a means of verification is to assume the burden of proof that the other side is not cheating. But since black boxes cannot provide 100% assurance, the true skeptic will never be satisfied that sufficient proof has in fact been given. It may, therefore, be much more in the interest of arriving at a treaty to recognize this inherent stability, and hence not to assume this burden of total verification at all, which the reliance on black boxes to enforce the treaty would imply.

The use of black boxes may indeed make more difficult some kinds of treaties. But it should facilitate the negotiation of others that would have been otherwise impossible. A camera pointing at a missile to ensure that a MIRV is not installed is just one example.

Other Requirements of Reconnaissance. The kind of reconnaissance discussed here is useful only to discover what is actually in existence in another country. To determine what is likely or what is possible requires entirely different techniques. Such techniques are particularly important because the knowledge of what is *possible*, in the absence of the knowledge of what is *likely*, is often what drives the arms race. One common practice now used for determining what is likely utilizes linear extrapolation from the present and the recent past. Since this takes no account of intentions, however, it has little or nothing to do with reality.

Performance of Anti-ballistic-missile Systems

B. Alexander

Introduction

The continuing debate on deployment of the Safeguard ABM system has brought many of the issues associated with ABM into public view. Much about ABM has been revealed in the many thoughtful reviews that have become available in recent years (Refs. 1–6). While the United States (and most probably, the USSR) continues its ABM-related advanced research, enough is now known to permit important conclusions to be drawn with reasonable confidence. Although political issues are vital to the ABM debate, they are not considered in this paper. The concern here is primarily technical in an attempt to answer the closely related questions, "How well would an ABM system work?" and, "How much would it cost?"

There are strongly held technical disagreements between the experts. The critics of ABM systems generally agree with Sakharov that the measure-countermeasure opportunities so heavily favor the offense as to create "technical and economic obstacles to an effective missile defense that, at the present time, are virtually insurmountable" (Ref. 7), although not all critics are as careful as Sakharov to qualify their conclusions. Garwin and Bethe believe a thin area defense to be so easily penetrated that "it is well within China's capabilities to do a good job at this without intensive testing or tremendous

Ben Alexander is Chairman of the Board of General Research Corporation.

sacrifice in payloads" (Ref. 1). The ABM enthusiasts disagree. Herzfeld, an advocate of "thin defenses," believes "that a system like Safeguard could defend the population of the United States against attacks from the CPR for many years" (Ref. 8). Brennan favors a "thick" ABM system because it "might change the postwar U.S. situation from one in which one-half the population was gone and recovery in any time period would be problematical to one in which perhaps ninety percent survived and economic recovery might be complete within five to ten years. This difference would be enormous" (Ref. 9).

The disagreements concern cost as well as performance. Chayes et al., while believing Safeguard would fail because of specific weaknesses of that system, seem to reject ABM systems in general because "the cost of overwhelming the system is significantly less than the cost of deploying it" (Ref. 10). Brennan (Ref. 11) asserts, to the contrary, that the offense/defense cost-ratio is near unity!

To understand these disagreements it is important first to distinguish between area and terminal missile defense. For example, Garwin and Bethe (Ref. 1) go into some detail in describing penetration aids which they say defeat an area ABM system. But against terminal ABM systems they suggest that the preferred offense options are evasion (i.e., attacking the undefended cities) or interceptor exhaustion (obtained by concentrating on a few of the defended cities). They seem to be saying that area ABM performance would be so poor as to make cost irrelevant, while terminal ABM would cost much more than the offense force which counters it.

Herbert York discusses penetration aids against terminal ABM defenses. "Will it work? By this question I mean: Will operational units be able to intercept enemy warheads accompanied by enemy penetration aids in an atmosphere of total astonishment and uncertainty?" He thus equates "working" to "successfully countering penetration aids," and concludes that "One must have serious reservations about whether these problems can ever be solved" (Ref. 3).

So it would seem that there are three main lines of technology-related criticism of ABM systems.

1. Area ABM defense won't work in the sense that readily available penetration aids can surely overwhelm it.
2. Terminal ABM defense can be easily and cheaply defeated by concentrating the attack, or circumvented by attacking undefended targets.
3. The technical problems of developing and deploying a reliable and responsive terminal ABM system which would work as designed at the scale of attacks to be expected are insoluble and are likely to remain so.

The criticisms posed above will be discussed in order. In the following section I concludes that while area ABM defense would clearly be only a minor impediment to an aggressive superpower, its effectiveness against a less powerful aggressor is likely to remain uncertain. Next, I address myself to the cost and effectiveness of terminal ABM defense and show that even if there were no technical problems, terminal ABM defense of population can be overcome at a small fraction of the cost of deploying it.* Terminal ABM defense of missile silos from a preemptive attack, however, would be comparable in cost to the offense. In the final section I briefly review ABM technology and argue that recent advances, particularly in radar and data processing, may have made a terminal ABM system with substantial performance technically feasible.

Area ABM Defense

With few exceptions, experts agree that area ABM defense is, and will with high probability continue to be, highly vulnerable to penetration aids. In the most comprehensive generally available review, Garwin and Bethe (Ref. 1) discuss penetration aids for countering area ABM defenses—including decoys, balloons, chaff, fragments, jammers, and the use of nuclear blackout. They argue authoritatively that penetration aids permit a technically versatile aggressor to nullify an area defense. Although, as Sakharov suggests, new technology could possibly change this, and just how large a penetration-aid-equipped missile force would be required is debatable, few differ with Herzfeld's assertion (Ref. 12) that area defense would have little effect on determined and powerful nations such as the United States or the USSR.

This unanimity, as we have seen, does not extend to a technically advanced area ABM defense

* U.S. officials have usually taken this position. The design of the Safeguard ABM system reflects the view that terminal defense of cities against large attacks is not feasible.

operating against attacks by lesser powers. Since the (presumed) lesser technical sophistication and smaller numbers of the offense make it more nearly on a par with the ABM system, political issues assume primary importance. The technical experts' disagreements often represent only a few years' difference in their guesses as to when some level of capability is likely to be reached or may reflect differing tacit assumptions regarding each side's strategic objectives. Disagreements on such questions will probably continue indefinitely and would be resolved by technical considerations only in the unlikely event of major breakthroughs.

Effectiveness and Cost of Terminal ABM Systems
Recognizing the various defense vulnerabilities, ABM engineers naturally adopt the concept of "balanced defense." That is, they attempt to shape their designs to deny the offense any clearly preferred modes of attack. This process is not perfect; but the possibility of a countermeasure invariably stimulates counter-countermeasures, and sometimes entails very major system changes. While the ABM developers may ignore a newly perceived problem or design it out of the "threat," under the pressure of deadlines or money limits, it often works the other way. The defense designers tend to overreact to penetration aids in order to achieve balance, thereby making the performance poorer than it could be (for the same money spent) against simpler attacks.

This suggests one way of getting at the "will it work" question. The most obvious way for an ABM system to fail is by exhaustion of its supply of interceptors—either because of a large number of attacking warheads (or decoys), or because interceptors are unreliable, so that many must be shot at each warhead. This defense failure mode (leakage and exhaustion) can be analyzed under the assumption that in all other respects the defense works perfectly. If the ABM design were balanced, the quality and number of interceptors would be determined by an allocation of defense costs between interceptors, radars, data processors, etc., such as to make it fail by "leakage and exhaustion" about as soon as it failed any other way. Therefore, by analyzing this failure mode, many of the technical characteristics of a balanced system can be inferred. This is a convenient starting point for discussion of cost ratios and evasive tactics as well as of technology.

The procedure in what follows will be to define reasonable estimates of the number of interceptors it takes to defend against "leakage and exhaustion," and then use this result to define the likely range of offense/defense cost ratios. This analysis will deal with the exchange between the offense and a single ABM "battery" defending one target. After that, the question of evasion and concentration against a widely deployed ABM system will be taken up.

Calculations of ABM exhaustion, and the related problem of doctrines and deployment for both offense strategic missiles and ABM batteries, are best handled by computer programs based on the Lagrangian multiplier methods originally developed by Everett (see Refs. 13, 14), but there are simple approaches which give reasonable results and can be approximated analytically. One, usually associated with R. C. Prim and W. T. Read, Jr., employs a defense doctrine of allocating interceptors in such a way as to make the offense's expectation of target damage equal for all warheads. It explicitly allows for interceptor unreliability, the effects of decoys, and the defense's estimate of the significance of physical damage.

As shown in Appendix 1, when a defense is designed to protect against a substantial attack, the approximate relation holds:

$$\frac{\text{Number of interceptors}}{\text{Number of warheads plus decoys}} = \frac{N}{M} = \frac{\ln (M\delta\eta/\Delta)}{\ln [1/(1 - P_I)]}, \quad (1)$$

where N = Number of interceptors needed to counter attack consisting of M objects; M = total number of warheads plus indiscriminable decoys; δ = fraction of remaining target destroyed by one warhead; η = fraction of objects that are warheads; Δ = fraction of target destroyed after the engagement; P_I = interceptor single shot kill probability.*

There are many subtle points that this model does not include such as the amount of knowledge available to offense and defense, differences between their objectives, etc., but these are not very important in this context.

* Note that in this model interceptors are *not* replaced if they fail.

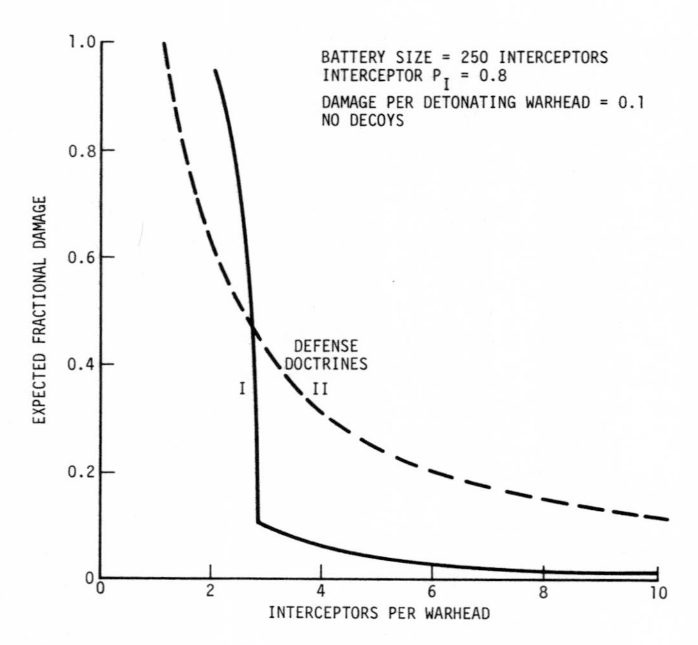

Figure 1
Expected target damage versus interceptors per warhead.

In the next few pages, a specific example is discussed in some detail. It will become apparent that the specific numbers used could be changed substantially without significantly altering the results. Figure 1 shows the outcome as a function of the number of arriving warheads. For this example, the defense has 250 interceptors, each with 0.8 single shot kill probability. Each warhead is assumed capable of destroying 0.1 of the target remaining. The two doctrines shown were the defense's choice: in I, a doctrine appropriate to defense of cities, the ABM

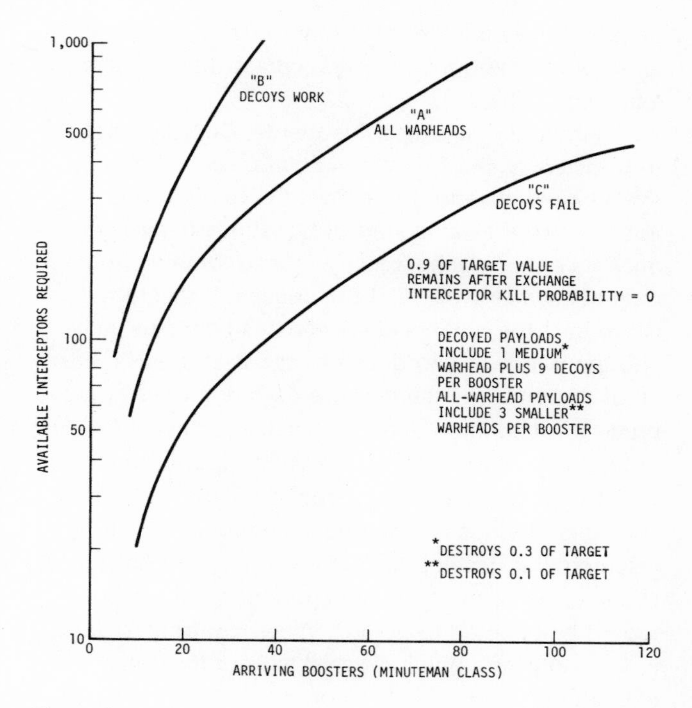

Figure 2
Interceptors required for effective ABM defense versus attack size.

system was used so as to exhaust its interceptors when 0.9 of the target remained undamaged, while in II, more applicable to missile site defense, it was designed to exhaust just as the target was destroyed.

Note that at more than 3 interceptors per warhead, the defense "wins," while at much less than 3 per warhead, the offense largely destroys the target. These conclusions apply to ABM batteries with roughly 250 interceptors, where the interceptor reliability is 0.8.

Figure 2 elaborates this theme. Curve A applies to boosters of the Minuteman class, each carrying three warheads and no decoys. Curves B and C apply when the same kind of booster is used with only a single (much larger) warhead, but carrying nine decoys as well. In B the decoys draw fire while in C the defense discriminates between them and the warhead and does not shoot at decoys. For all of these curves the defense uses doctrines appropriate to population defense, similar to that of Curve I in Figure 1. The contours shown in Figure 2 refer to 0.1 expected fractional target damage.

It is obvious that the defense would aim its research and development toward getting on the C curve, while the offense penetration aid program would be directed toward B. When confronted with a defense thoroughly prepared to counter penetration aids, the offense is likely to be driven toward use of multiple warheads (and hence to the A curve) by the fear that decoys would fail, forcing it to operate on the highly disadvantageous C curve. To the extent that this is true, it would seem that, for terminal ABM defense in the size range of 250 (0.8 reliable) interceptors per battery, three interceptors per reentry vehicle as shown in Figure 1 would meet the defense objectives, assuming that in all other respects the ABM worked perfectly.

The per-battery cost ratio, for the cases we have been discussing, can now simply be inferred from available cost data. Dr. John Foster, Director of Defense Research and Engineering, wrote that "estimates of the cost of an interceptor, including its assigned fraction of the radar and other systems cost, have varied between $2.5 million and $7 million" (Ref. 15). He also said that the present cost to the United States, and probably to the Soviet Union, of an offensive reentry vehicle is in excess of $10 million, although this may be reduced to about $3 million. These figures (using three interceptors per reentry vehicle in a terminal ABM system) lead to offense/defense cost ratios between 1:0.8 and 1:7. Foster also suggests that offense costs will never go much lower, so the real ratio may lie near the lower edge of the range if ABM development goes into large-scale production (as would be required for a thick defense). Thus, while the cost ratio argument, viewed this way, favors the offense, the margin could approach 1:2 and conceivably even 1:1.

The above applies to a single battery with 250 interceptors. As the forces on both sides increase, the ABM system must improve its capability (relatively) in order to keep the damage constant. This is shown by Equation 1, where the interceptor/warhead ratio increases logarithmically with the number of warheads for constant damage. However, over a wide range of offense and defense levels around the 250-interceptor battery of Figure 1, the logarithmic term is not too significant, in spite of the interceptor single-shot kill probability being only 0.8 and the defense doctrine that of maintaining 90% of target value up to the point of interceptor exhaustion. For example, as the number of arriving warheads increases from 90 to 900, the number of interceptors needed to hold expected damage constant increases from 250 to 3,800. This increase of interceptor/warhead ratio from 2.8 to 4.2 is nót enough to affect the cost ratios violently.

The problem of evasion of defenses and concentration of the attack that Garwin and Bethe describe as the normal mode of failure of terminal ABM defenses can be treated by considering "over-all," as opposed to "per-battery," cost ratios. For this strategy to make sense, the offense's objective must be aggregated damage to population, or floor space, or missile silos, rather than a set of

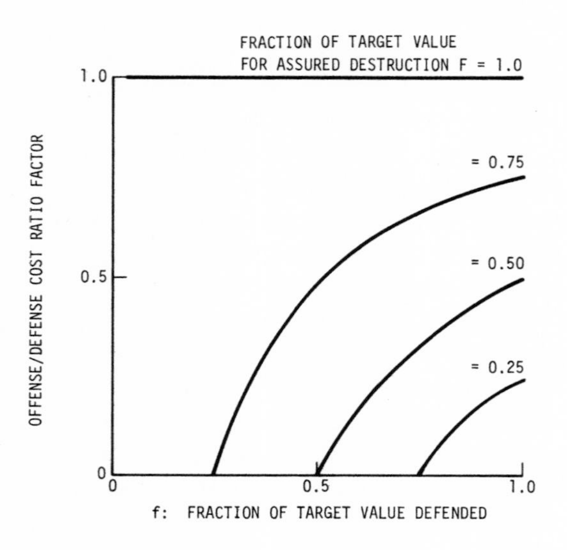

Figure 3
Cost ratios versus fraction defended.

specific targets, particularly if they are known to the defense. In what follows the offense objective is taken to be a stated fraction of the enemy's urban population.

Appendix 2 develops an "offense/defense cost ratio factor" as a function of the fraction of the target value that is defended and the fraction of the target value the offense elects to threaten with "assured destruction" notwithstanding. When the offense has large numbers of warheads and, therefore, can "overkill" many times over,* much of the urban population *not* defended by the ABM system is threatened with assured destruction at a cost that may be considered negligible relative to the cost of the whole force. The "cost ratio factor" is the factor by which the "per-battery cost ratio" must be multiplied to obtain the "overall offense/defense cost ratio." Figure 3 plots this factor as a function of the "thickness" of the ABM deployment, that is, the fraction of the target system defended. The cost ratio factor is shown for a range of values of the offense's "assured destruction" objective.

It is apparent that the "thinner" the ABM deployment the more the cost ratio factor favors the offense. This is at the heart of many objections to terminal ABM defense. Notice that even if the offense insists on threatening as much as half the value of the target system, and the defense is thick enough to have batteries covering three-quarters of

* York (Ref. 3) shows that the United States and the USSR could each devastate the other's fifty largest cities with a few percent of their post-MIRV warhead inventory.

all the target value, the cost ratio factor is one-third. If the offense slightly lowered its sights and was satisfied to threaten destruction to one-third (as opposed to one-half) of the opponent's target value, the cost ratio factor would fall to one-ninth. Since this factor multiplies the per-battery cost ratio (which is not likely to be higher than 1:2), for these cases the overall offense/defense cost ratio would be less than 1:6 and 1:18, respectively.

Not only will the cost ratio be disadvantageous, but the absolute cost of terminal ABM defense will be high because of the granularity of ABM costs and the number of quite small cities that constitute the "second half" of urban targets in advanced countries. Since ABM radar and data processing costs depend more on the elements involved in an engagement (including the rates and the interactions provided for) than on just the number of interceptions, the cost per interceptor is excessive if the batteries become too small. However, if the fraction of the target value that must be defended is more than, say, fifty percent (in order to avoid the very small cost ratio factors shown in Figure 3), many small cities must be defended.

Appendix 3 discusses this problem using highly simplified but reasonable models. It is assumed that a city's value is proportional to its population and that population varies inversely with the rank of the city, in order of population. ABM granularity is represented by defining a minimum battery size. ABM interceptors are allocated to defended cities in numbers proportional to their

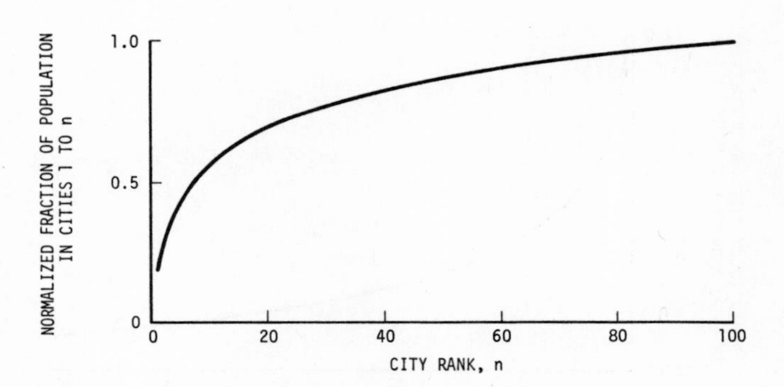

Figure 4
Fraction of population in cities 1 to *n*, inclusive, if city population is inverse to rank.

value.* The resulting approximate relationship is

$$\frac{\text{Interceptors at all cities}}{\text{Interceptors at least city defended}} = T^f \ln T^f, \quad (2)$$

where T = the total number of cities in the target system, and f = the fraction of total value defended.

Figures 4 and 5 show a numerical example. From Figure 4 it can be seen that if the ABM system were to be deployed to defend 75% of the population of the first 100 cities, the least defended city would be the 30th. Figure 5 shows that the 30th city has about 1% of the population of the top 30 cities. Hence, if the minimum ABM battery consists of, say, 50 interceptors, the minimum national deployment would be 5,000. Using the middle of Foster's range of costs, $5 million per interceptor, for the whole

* The defense generally worsens its position by defending other than proportional to value; see Appendix 2.

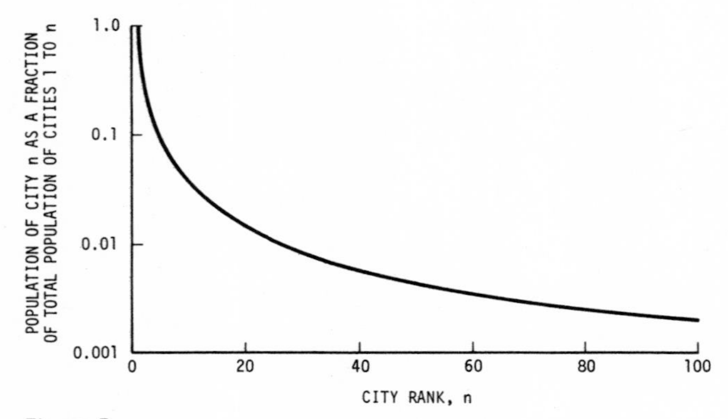

Figure 5
Population of the *n*th city as a fraction of the population of cities 1 to *n*, if city population is inverse to rank.

ABM system would lead to a national deployment cost of $25 billion. This is the minimum cost for a defense which is thick enough, as we have seen, to hold the overall offense/defense cost ratio to between 1:6 and 1:18. Hence, the offense cost for neutralizing this defense ($25 billion) would be no more than $4 billion and possibly $1.5 billion. It can be seen that at a $50 billion ABM investment the defense is not much better off.

This line of argument leads to the inescapable conclusions that a "thin" or even moderately "thick" terminal ABM defense of cities is self-defeating against a superpower determined to maintain a posture of deterrence by keeping a fraction of its opponent's population at risk. At a low investment the overall offense/defense cost ratio is essentially zero, while at a defense investment as high as $25 billion the cost ratio is only 0.1 or 0.2. For the same reason, terminal ABM defense against lesser powers is likely to be of no use whatsoever, as their strategic objective would be to threaten only a small fraction of the population.

One exception to the above occurs in the case of a few unique or symbolic cities. It might be the offense's objective to destroy them all ($F = 1.0$) and the defense's to save them ($f = 1.0$). If this situation were to exist the cost-ratio factor, as shown in Figure 3, would be unity.

The possibility of terminal ABM defense of cities naturally suggests its use for the defense of military targets, especially for defending fixed missile silos from a possible preemptive attack. The cost-ratio arguments which discourage terminal ABM defense of populations are less applicable since, considering the overkill potential, the first strike objective must be a very large fraction of the other side's strategic force ($F \sim 1$). Moreover, the arguments leading to very high ABM costs because of the decreasing value of smaller cities do not apply. Furthermore, it may be possible to build a silo-defense ABM system somewhat more cheaply than one for city defense because of the hardness of the targets.

There is, in addition, the problem of preferential defense. Since the defense very likely is quite willing to sacrifice a fraction of its missiles if it is attacked, a defense which covers ten silos, say, may elect to concentrate its firepower to save one or two of its choosing. This forces the offense to attack with more warheads than interceptors in the ratio of the silos attacked to those defended. Since this could be a factor of five or ten in favor of the defense, the overall cost ratio might turn out to favor the defense quite substantially. To counter this, the offense is likely to respond by attacking the ABM radar, while the defense might respond to the response by replicating its radars or by other means.

All in all, the defense of missile silos cannot be rejected on cost-ratio grounds and there are quite obviously a number of strategic situations in which it might seem attractive.

Technology of Terminal ABM Defense

I now address myself to technology. Given the hypothetical 250-interceptor terminal ABM battery, is it, or will it become, technically feasible to achieve the balanced performance assumed, at about the cost estimated?

The leakage and exhaustion model used for the cost and effectiveness discussion does not allow for performance deterioration due to concentration of the attack in time or space. Since a terminal ABM system would use the atmosphere (Ref. 1) to eliminate chaff and light decoys, the defense must engage objects during the last 30–60 seconds of flight. There is no fundamental reason why the offense cannot fire most of its missiles to land within a much shorter time. As a consequence, the ABM system must be prepared to cope with near-simultaneous arrival of the attacking boosters and must be able to conduct virtually all operations entailed in 250 interceptions in, possibly, 30 seconds.

It is not practical here to even try to identify all of the major operations the defense must perform during this period. This would require constructing and debugging logic diagrams with literally hundreds of paths to ensure that the needed information is gathered, processed, and acted upon properly at each decision point. However, as a minimum the following sequence of major operations could be necessary within the final 30 seconds:

1. Radar acquires and tracks up to 300 objects.
2. Data is collected on each of the 300 objects and processed to set priority (i.e., likelihood of being a warhead, yield, "value" of probable impact point, etc.).
3. 250 interceptor targets are assigned, launch decisions made, and trajectories selected.
4. 250 interceptors are launched, tracked, guided and burst so as to destroy their targets without damage to other interceptors in flight or to radar visibility.

Certain technical implications are apparent.

The first, and perhaps most important, is that the defense cannot count on distributing its effort uniformly over the final 30 seconds of reentry. The 30–60-second time window (dependent largely upon reentry angle) sets the maximum time in which the defense may operate. Since ABM interceptors as a rule use thermonuclear explosives, opaque fireballs from earlier interceptions tend to progressively compress the engagement. Garwin and Bethe (Ref. 1) show that an explosion of only 10 kilotons produces a fireball 1 km in diameter at an altitude of 30 km, and 300 m in diameter at 10-km altitude. If the first one hundred defensive bursts occurred at or above 10-km altitude, they would occlude at least ten percent of the remaining objects aimed at a city 10 km in diameter. As a consequence, advances in radar sensitivity, resolution, and subclutter visibility, however great, would not overcome the need to search large solid angles at high data rate to detect targets which may have been shadowed by blackout and appear first at low altitudes. The search must be rapid (perhaps each second) and cover a large solid angle (perhaps 1 steradian). Moreover, the low leakage rates implied by the "balanced defense" concept would require an expected penetration of only 1 warhead out of 100 arriving. A single radar is likely to be occluded too frequently to assure this, so two separated radars contributing to a common target file are probably necessary.

Such radars—functionally the missile site radar (MSR) described by Chayes et al. (Ref. 10)—would have to seek out and track the incoming warheads and guide the defending missiles to the point of intercept. Thus, in the search function they must operate out to 150-km range (20–30 seconds before missile impact) while being sufficiently accurate to make targeting decisions rapidly and still cover the field of view in the order of one second.

The literature is replete with descriptions of electronically scanned radars (Ref. 16). These can have a wide variety of waveforms and beam shapes, which may be computer selected on a pulse-to-pulse basis. An MSR to work with the hypothetical 250-

interceptor battery would be a radar of this general type. Having accepted this, its characteristics do not look too formidable, although poor reliability and high cost could result from the very large number of components involved.

For example, a radar with a 300-km range and 3° beamwidth (15-km resolution and perhaps 0.3-km cross-range accuracy at 300-km range) can search one steradian per second at full range with a single beam and simple waveform. Compared to a long-range search radar such as PAR, its power-aperture requirements are lower by perhaps a factor of 10^3* (although in other respects it may be more demanding). Moreover, with solid-state distributed power generation available (Ref. 16), sensitivity is not likely to be too important a performance limit. Its required accuracy for tracking missiles and interceptors, because of the use of thermonuclear explosives in the interceptor, is not excessive.

The data-processing loads are sure to be heavy but not likely to limit performance of the hypothetical 250-interceptor battery nor dominate its cost, although there is room for doubts about the software.

A possible analogy to an ABM system in terms of data-processing complexity is automatic air traffic control. A recent comprehensive air traffic control study (Refs. 17, 18) considered the feasibility of developing a terminal-area air traffic control data processor capable of simultaneously handling 4,200 aircraft, including 1,500 under close (IFR) control.

* Assuming a range of 3,000 km and angular coverage rate of 1/10 steradian per second for PAR.

The cooperative ranging subsystem required an average rate of 0.3 data points per second for each of 4,200 aircraft, and the data processor had to maintain tracks on each. These were to be correlated with the outputs of 5 search radars, each with a four-second scan and capable of generating up to 100 new data points per second. The data processor would schedule and control takeoffs and landings (almost one per second on the average), perform automatic conflict detection and resolution functions for all 4,200 aircraft each 20 seconds, and coordinate with the enroute system to achieve national flow control.

The air traffic control (ATC) processor may well be more complicated than the ABM's. Whereas the ABM system might acquire 300 objects per second and correlate with a file of about the same size, the ATC processor acquires 500 radar targets and 1,260 cooperative targets per second and correlates with a file holding 4,200 aircraft. While the ABM system might be guiding 100 interceptors simultaneously, each at a data rate of perhaps 10 per second, this is probably less of a load than searching out conflicts for 105 aircraft (4,200 × 1/20 × 1/2) each second with each of 4,200 others. Both systems, of course, have many additional computations to perform.

The ATC study sized the computer at 10 million instructions per second (MIPS) ideally, and 100 MIPS after allowing for practical—and hence inefficient—programming. Reliability was specifically "designed in" by replication of key units and by

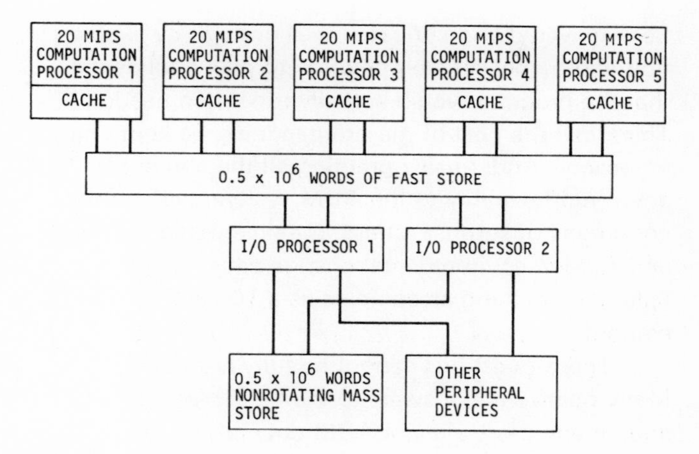

Figure 6
Maximum terminal air traffic control processing (1990 projection).

including "caches" at each processor to avoid loss of data when a failure occurs. Figure 6, taken from Ref. 17, shows a simplified overall diagram.

The study team, which included representatives of the major American computer manufacturers, reviewed the state of the art and concluded that all components required could be commercially available by about 1975. The estimated manufacturer's selling price for the complete computer, less peripherals, fell in the range $12.6–25.1 million (Ref. 17).

There is less known about the software, and four questions are frequently raised. First, can logic be developed and tested sufficiently well in simulated engagements so that even with perfect programming the ABM system could run through the millions of steps necessary? Second, can programming errors be discovered and corrected well enough by simulation and by peacetime operation to give confidence that the program would actually run when needed? Third, can the cost of the programming be kept within reason? And, finally, granting all the above are answered favorably to the ABM system, will not the conditions of surprise, stress, possible partial damage, and sudden environmental change cause major failure in anything as complex as a 100 MIPS computer?

These questions cannot be fully answered. Many believe the answers to the first three are "yes," and, in any case, a major ABM data processor development program would provide most of the answers. The fourth question is much more difficult to assess. While it applies to other complex military systems as well, it is most relevant to the ABM system, particularly if designed for terminal defense. It is not provable, either positively or negatively, and will probably remain a matter of judgment with opinions strongly held on both sides.

Interceptors such as Sprint cost somewhat less, including warheads, than the larger Spartan for which Foster assigned a total system cost of $2.5–7 million per interceptor. Sprint is a modern, twenty-five-foot long, two-stage, solid-fueled missile equipped with a thermonuclear warhead in the kiloton range, used for terminal defense (Ref. 20). Its conical forebody shape suggests that it reaches high speed in the lower atmosphere, and its conical shape overall implies very high acceleration. Such performance is consistent with the response time that has been discussed. Interceptor accuracy and reliability depend upon design details, but suggestions that 0.8 kill probability is reasonable have been published (Ref. 21). This could be 0.9 or even higher without changing the cost ratios appreciably.

With respect to all of the above—interceptors, data processors, radars, and of course the communication network between them—there are the problems of equipment reliability, under both peaceful and wartime conditions. This is an extremely complex subject, due to the hundreds of millions of components involved and the very many paths by which they interact. While no one can say for certain, the case can be made that (1) recent and continuing advances in materials and techniques (in particular semiconductors for large-scale integration of electronics) and (2) "fallout" from the missile and manned space programs in the knowledge of designing for reliability make it quite likely that ABM batteries can be made to operate in peacetime with an acceptable failure rate. Thus, the reliability problem is analogous to the software problem in that very much can be learned during development programs. The wartime performance is far less certain, dependent as it is on details of the attack, nuclear-burst phenomena which cannot easily be experimented with, and the many ways damage-caused failures can propagate through the system.

The foregoing addressed the major elements of technology involved in the performance of terminal ABM defense, assuming the attack consisted of perhaps 100 warheads concentrated in time. The conclusion seems warranted that by using phased-array radars, large data processors (of the order of 10^8 instructions per second), and high-velocity, high-acceleration nuclear interceptors, terminal defense may turn out to work well enough so that leakage and exhaustion adequately describe its eventual failure. The major uncertainties (assuming a successful development program) have to do with the hardware and software functioning properly under the stress of a nuclear attack.

Turning to penetration aids, a number of those listed by Garwin and Bethe (Ref. 1) for use against area defense have some applicability against terminal systems, although ABM countermeasures are possible in each case. It is impossible to know for sure how a large-scale technological arms race between penetration aids and ABM countermeasures would turn out. There are two reasons for believing the outcome would not be much worse for the defense than the leakage and exhaustion results already quoted. First, the defense would include phased-array radars, supercomputers, and Sprint-type interceptors to cope with a coordinated multiple-warhead attack. It would have, therefore, great reserves of performance, especially in radar and data processing, which it almost certainly would apply to countering penetration aids. Moreover, because the details of this capability are concealed, an offense is inhibited from using penetration aids that exploit specific weak points of the ABM design. The second, and related, reason is that multiple warheads are inherently reliable and relatively inexpensive. It is doubtful that either side would bet too heavily that practical penetration aids better than multiple warheads could be developed for use against terminal ABM defenses.

Appendix 1 Firing Doctrines

The Prim–Read firing doctrine is close to the optimum if the offense uses a uniform mix of warheads and decoys. With this allocation the defense maintains the expected fraction of target damage per incoming object constant. Using an "exponential" damage model (in which each bomb destroys the same fraction of the *remaining* target value), the constant expected damage per object, λ, is

$$\lambda = V_i(1 - P_I)^{\bar{n}_i}\,\delta\,\eta, \tag{1}$$

where λ = value expected lost to the defense per penetrating warhead; V_i = remaining target value before the ith attacking object; P_I = probability a launched interceptor kills a warhead or decoy; $\bar{n}_i$ = average number of interceptors launched against the ith object; δ = expected fraction of the remaining target destroyed by a warhead (or in the small-hard-target case, the probability of a penetrating warhead destroying the target); and η = fraction of objects that are warheads (decoys assumed harmless but indiscriminable). Hence,

$$\left(\frac{1}{1 - P_I}\right)^{n_i} = \left(\frac{V_0}{\lambda} + 1 - i\right)\delta\,\eta, \tag{2}$$

where V_0 is the original value of the target, i.e.,

$$V_i = V_0 - \lambda\,(i - 1).$$

If the defense has N interceptors and is exhausted after the Mth object,*

$$\prod_{i=1}^{M}\left(\frac{1}{1-P_I}\right)^{\bar{n}_i} = \left(\frac{1}{1-P_I}\right)^{N}$$

$$= (\delta\,\eta)^{M}\prod_{i=1}^{M}\left(\frac{V_0}{\lambda}+1-i\right)$$

$$= (\delta\,\eta)^{M}\left(\frac{V_0}{\lambda}\right)!\left[\left(\frac{V_0}{\lambda}-M\right)!\right]^{-1}. \qquad (3)$$

For an attack which leaves the fraction Δ of the target destroyed after M objects, i.e., $V_0/\lambda = M/\Delta$, Equation 3 can be expressed as

$$N \ln\left(\frac{1}{1-P_I}\right) = M \ln(\delta\,\eta) + \ln\left(\frac{M}{\Delta}\right)! - \ln\left(\frac{M}{\Delta}-M\right)!. \qquad (4)$$

Originally the Prim–Read doctrine was formulated to leave the expected target value zero when the interceptors were exhausted. In a sense this "bought back" the most damage per interceptor. For this case $\Delta = 1$ and Equation 4 can be expressed as

$$N \ln\left(\frac{1}{1-P_I}\right) = M \ln(\delta\,\eta) + \ln M!. \qquad (5)$$

* This is obtained by setting $N = \sum \bar{n}_i$, which is not strictly correct as the defense can only approximate $\bar{n}_i$ by statistically selecting salvos sizes of the integers above and below $\bar{n}_i$. This causes an average overexpenditure of interceptors per salvo that varies from 0.2 at $P_I = 0.95$ to 0.1 at $P_I = 0.67$. In the range of 3 interceptors per salvo, at $P_I = 0.8$, the approximation results in underestimating the interceptors required by 4%.

By Stirling's formula,

$$\ln M! = M \ln M \left[1 - \frac{1}{\ln M} + \frac{1}{2M} + \frac{\ln 2\pi}{2M \ln M} + \frac{\theta}{12M^2 \ln M}\right],$$

where $0 < \theta < 1$. For $M > 30$ the third term in the bracket is less than 1/60, the fourth term is less than 0.01, and the term in θ is less than 10^{-4}. Hence, to a close approximation

$$\ln M! = M \ln M - M \qquad (M > 30). \tag{6}$$

For doctrines designed to leave the target largely intact up to the point of interceptor exhaustion, $\Delta < 1$ and Stirling's formula must be applied to Equation 4 instead of Equation 5. This yields, after some reduction,

$$\ln\left(\frac{M}{\Delta}\right)! - \ln\left(\frac{M}{\Delta} - M\right)! = M \ln \frac{M}{\Delta} - M + \left(\frac{M}{\Delta} - M + \frac{1}{2}\right) \times [-\ln(1 - \Delta)] + \frac{\theta_1}{12M/\Delta} - \frac{\theta_2}{12M(1 - \Delta)/\Delta}. \tag{7}$$

But $-\ln(1 - \Delta) = \Delta + \frac{1}{2}\Delta^2 + \frac{1}{3}\Delta^3 + \ldots$, and for $M > 30$ the terms in θ are negligible unless $\Delta \sim 1$, in which case the previous derivation shows the θ_1 term is negligible and the θ_2 term does not appear. Therefore,

$$\ln\left(\frac{M}{\Delta}\right)! - \ln\left(\frac{M}{\Delta} - M\right)!$$
$$\sim M \ln \frac{M}{\Delta} - \frac{\Delta}{2}(M - 1) - \frac{\Delta^2}{6}\left(M - \frac{3}{2}\right)$$
$$- \frac{\Delta^3}{12}(M - 2), \ldots,$$

and

$$\ln\left(\frac{M}{\Delta}\right)! - \ln\left(\frac{M}{\Delta} - M\right)!$$
$$= \left[M \ln \frac{M}{\Delta} - \frac{M\Delta}{2}\right](1 + \varepsilon) \tag{8}$$

where $\varepsilon < 0.05$ for $M > 30$ and $\Delta < 0.75$, and rapidly diminishes for smaller Δ. Therefore within acceptable accuracy, from Equations 4, 5, 6, and 8, as long as $M > 30$:

$$N = \frac{M \ln M\delta\eta - M}{\ln [1/(1 - P_I)]}, \qquad \Delta \sim 1, \tag{9a}$$

$$N = \frac{M \ln (M\delta\eta/\Delta) - (M\Delta/2)}{\ln [1/(1 - P_I)]}, \qquad \Delta < 0.7, \tag{9b}$$

and, for the defense which seeks to survive a substantial attack up to the point of interceptor exhaustion, i.e., $M \gg 30$, $\Delta \ll 1$

$$N \sim \frac{M \ln (M\delta\eta/\Delta)}{\ln [1/(1 - P_I)]}. \tag{10}$$

Appendix 2 Proportional-to-value Defense

If the offense's objective is merely to assure the destruction of a fraction F of the defense's target value, and the defense elects to defend a fraction $f \leq 1$ proportional to value, the offense naturally attacks the undefended fraction first. Hence, the additional cost the offense must incur because of the ABM is proportional to the part of the defended fraction it must attack, $F - (1 - f)$. The defense cost is clearly proportional to f. Therefore, the offense cost is below what it would be if all the value attacked were defended and the defense cost is above what it would be if it defended only the value attacked. The factor by which the per-battery offense/defense cost ratio is reduced by the offense's ability to evade and concentrate remains at zero unless $f > 1 - F$ and for larger values of f is obviously

$$\frac{F - (1 - f)}{f}.$$

This is plotted in Figure 3 in the text.

If the defense elects to defend other than proportional to value it is even worse off. Consider a target system T made up of two parts T_1 and T_2, with fractions f_1 and f_2 defended. Suppose T_1 were defended with λ_1 interceptors per unit value and T_2 with λ_2, with $\lambda_1 < \lambda_2$. The offense would first attack the undefended portions, then the defended part of T_1, and finally the defended part of T_2. If

$$FT < (1 - f_1)\, T_1 + (1 - f_2)\, T_2 + f_1 T_1,$$

clearly, having $\lambda_1 \neq \lambda_2$ makes the offense less costly. Even if *FT* were larger and some targets defended at the λ_2 level must be engaged, there would be more unused interceptors (and hence lower offense costs) for $\lambda_1 \neq \lambda_2$.

Appendix 3 Allocation of Interceptors Among Defended Cities

If there are $T \gg 1$ possible targets with values (in order from the most valuable) $v_1, v_2, \ldots,$ and they are defended proportional to value up to some fraction of the total value, the number of interceptors defending the jth target, I_j, is

$$I_j = \lambda v_j \text{ for } \sum_{i=1}^{j} v_i < fV, \qquad (1)$$

$$I_j = 0 \text{ otherwise,}$$

where I_j = number of interceptors defending the jth target; λ = constant number of interceptors per unit defended value; v_i = value of the ith target, with targets ordered inversely to value; f = fraction of total target value defended; V = total target value; and N = total number of interceptors.

The total number of interceptors N is related to the total defended value by

$$N = \lambda \sum_{\substack{\text{defended} \\ \text{target}}} v_i = \lambda f V. \qquad (2)$$

Because of radar and other fixed-facility costs there is a minimum practical ABM battery size—measured in number of interceptors. If a target system

(T targets) is made up of unequal-valued targets, and if targets are defended proportional to value, the fraction of the target system to be defended and the minimum battery size determine the total number of ABM interceptors and, therefore, the total ABM cost.

If the value of the least well defended target is v_μ, and the number of interceptors defending it is $I_\mu = \lambda v_\mu$, Equation 2 can be expressed as

$$\frac{N}{I_\mu} = \frac{fV}{v_\mu}, \qquad (2a)$$

where I_μ is the number of interceptors in a minimum battery. City value (population, floor space) in advanced countries varies roughly inversely with city rank. Since the fraction of the total number of interceptors at the jth city is merely j's fraction of the population in cities its own size and larger, the N/I_μ ratio for the value-inverse-with-rank model depends only on the fraction of the target system defended. Figures 4 and 5 in the text show this. From Fig. 4 one determines the least defended city on the basis of fraction of population to be defended and then Fig. 5 gives the I_μ/N ratio.

The analytic relationship is easily seen. For the value-inverse-with-rank model, the defended value fV is equal to $\mu v_\mu \sum_{i=1}^{\mu} (1/i)$, and Equation 2a becomes

$$\frac{N}{I_\mu} = \mu \sum_{i=1}^{\mu} \frac{1}{i}$$

or, to the approximation $\sum_{i=1}^{\mu} (1/i) = \ln \mu$,

$$\frac{N}{I_\mu} = \mu \ln \mu. \qquad (3)$$

In terms of f, the fraction of the targets defended, and the total number of targets T (defended and not defended), since $\mu = T^f$,

$$\frac{N}{I_\mu} = T^f \ln T^f. \tag{3a}$$

Thus, cost rises faster than exponentially with the fraction defended.

References

1. R. L. Garwin and H. A. Bethe, "Anti-Ballistic-Missile Systems," *Scientific American*, vol. 218, no. 3 (March 1968).

2. G. W. Rathjens, "The Dynamics of the Arms Race," *Scientific American*, vol. 220, no. 4 (April 1969).

3. H. F. York, "Military Technology and National Security," *Scientific American*, vol. 221, no. 2 (August 1969).

4. C. M. Herzfeld, "Missile Defense: Can It Work?," *Why ABM?*, Pergamon Press, New York, 1969, p. 25.

5. W. R. Kintner et al., *Safeguard: Why the ABM Makes Sense*, Hawthorn Books, New York, 1969.

6. A. Chayes and J. B. Wiesner, Editors, *ABM: An Evaluation of the Decision to Deploy an Antiballistic Missile System*, Signet Books, New York, 1969.

7. A. D. Sakharov, *Progress, Coexistence, and Intellectual Freedom*, W. W. Norton, New York, 1968, p. 35.

8. Ref. 4, p. 23.

9. D. G. Brennan, "The Case for Population Defense," *Why ABM?*, Pergamon Press, New York, 1969, p. 94.

10. A. Chayes et al., "ABM Deployment: What's Wrong With It?," *ABM: An Evaluation of the Decision to Deploy an Antiballistic Missile System,* Signet Books, New York, 1969, chapter 1.

11. Ref. 9, p. 95.

12. Ref. 4, p. 39.

13. H. Everett III, "Generalized Lagrange Multiplier Method for Solving Problems of Optimum Allocation of Resources," *Operations Research,* May–June 1963, p. 399.

14. R. J. Galiano, *Defense Models VI, Strategies of Preferential Defense in Combination with Fixed Terminal Defense,* Lambda Corporation, Arlington, Virginia, March 1968.

15. J. S. Foster, Jr., "The Safeguard Decision Before Congress," *Safeguard: Why the ABM Makes Sense,* Hawthorn Books, New York, 1969.

16. P. J. Klass, "Advances Speed Phased Array Use," *Aviation Week & Space Technology*, June 22, 1970, p. 169.

17. *Report of the Department of Transportation Air Traffic Control Advisory Committee,* Vol. 2, appendix D, U.S. Government Printing Office, December 1969, p. 237.

18. N. A. Blake and J. C. Nelson, "A Projection of Future ATC Data Processing Requirements," *Proceedings of the IEEE*, vol. 58, no. 3 (March 1970), p. 391.

19. Ref. 16, p. 239.

20. Ref. 2, p. 22.

21. Ref. 4, p. 35.

Summary of Discussion

The Safeguard ABM System. It was pointed out that an optimum hard-point ABM defense would probably look very little like the present Safeguard system. Because of the technical history of Sentinal-Safeguard, the silo defense part of Safeguard is by no means optimized as either a minimum-cost or maximum-effective solution. Rather, the missile site radar, a very expensive and relatively soft item, is a very attractive target at which the attacker can aim his missiles.

Offense/Defense Cost Ratios. The offense/defense cost ratio may not be as favorable to the defense as was suggested, even for an optimum silo defense system, if the response of the offense is to employ increasingly more warheads per missile. Although MIRVing increases the cost of the missile, since only one booster is used for many warheads, the cost per warhead would decrease. As a consequence the offense/defense cost ratio is shifted favorably toward the offense by the deployment of MIRVs.

Even if the offense/defense cost ratio could be made to come out favorable to the defense, that in itself is not sufficient reason to build an ABM system. The ABM cost should first be compared to the cost of other methods of providing a secure deterrent.

Unambiguity of Silo Defense. The question was raised whether an ABM system of the type outlined for silo defense would be seen as unambiguously useful for that purpose only. Some aspects of the system, such as low-altitude, short-range interception and fairly low-power radars might contribute to this kind of restriction on the system's applicability. In response, it was pointed out that, although such an ABM system might have lower performance as a city defense than could otherwise be designed, it could still be used for the latter purpose and could appear to have that mission built into it.

Future Directions. Computer technology is progressing rapidly enough to meet proposed ABM needs. The terminal ABM system that was described is probably reaching a point of technological stability where the next big advances, if there are any, would be in cost reduction and in improvement of reliability and ease of operation.

Mid-course Interception of ICBMs. The mid-course region of ICBM flight is where the missile moves in free-fall under the influence of gravity only. Since there is about ten minutes' warning at this stage, even from the impact point, an ABM system could be developed which would cover a very large area. For example, a BMEWS radar tied to an ICBM-type interceptor could defend the whole United States. A radar like the Dog House radar outside Moscow, coupled with an ICBM interceptor, could provide a very large coverage of European Russia.

However, there are major sensor problems of target selection and tracking in the presence of a responsive offense, and in the environment produced by the nuclear explosions of previous boosters. The offense can employ balloon decoys which look like reentry vehicles and chaff, i.e., clouds of radar reflectors which conceal the reentry vehicles. The only observable available to the defense radars is the trajectory of the vehicle, i.e., where it is going and where it has come from, and its cross-section history. If the reentry vehicle is not going to an area which is to be defended, it may be ignored. A combination of balloons which show the same radar cross sections as real reentry vehicles and chaff could cause the defense to rapidly exhaust its supply of interceptors by shooting at everything.

A possible development, which could enhance the feasibility of midcourse interception, would be the deployment of a small sensor, either placed in a satellite or ready to be fired up toward the oncoming threat. This sensor could observe the threat at close range, thereby diluting the chaff and permitting an inspection at a wide range of aspect angles. This would make the offense's task of producing decoys which appear to be reentry vehicles much more difficult. It may not be possible to show that such a sensor could be produced, but on the other hand, neither may it be possible to prove that it could not be produced. Hence, the worst-case analyst will eventually assume that it exists.

Ocean Technology

V. C. Anderson

Ocean technology—what an innocent name for the mélange of disciplines it encompasses. The very name itself is new, having been applied in the last few years to a vague subset of the omnibus term oceanography. However, newness in ocean technology is rarely a unique discovery giving rise to concepts initiated there. Instead, newness in ocean technology usually consists of the adaptation of some aspect of a method, idea, or technique developed in other fields to solve problems beset by the peculiarities and vagaries of the ocean environment. Oceanography and its companion, ocean technology, have classically been fields in which problems abound in not only the variety to be found, but also in the sheer numbers that occur because of the nearly limitless extent of the world's oceans. While the problems are abundant, the solutions are few and difficult to come by. Solutions are few because the ocean is a complex and hostile environment for most of the technology that we use on land today, and because, relatively speaking, the workers in these fields are few in number. Thus we find that many examples of new ocean technology, new in our narrow sense of that technology which can be brought to bear on the advancement of military systems, may encompass techniques which in reality are quite old and well known. Many concepts, many techniques have been

Dr. Victor C. Anderson is Acting Director of the Marine Physics Laboratory of the Scripps Institution of Oceanography in San Diego, California.

in existence for a great number of years, but the incentive and support generated for the exploitation of these concepts and techniques have been limited to the efforts of a handful of men.

In response to the topic of this conference, I have singled out a few items from this class of new ocean technology which I feel might have a particular impact on the arms race. The ocean itself conveniently breaks these items into three categories: the intra-medium category, which relates to the inner space of the ocean medium itself, and two boundary categories, one at the surface and the other at the ocean's floor. If the emphasis on acoustic technology seems to be high in the list which follows it is a direct result of the prejudice which I am sure is built in on my part toward the special impact of acoustics in the case of military applications.

Let us consider the intramedium category and start off by asking what's new in acoustics that wasn't in the bag of tricks 10 years ago. To set the stage I should point out that there are a number of aspects of acoustics that are important when talking about the performance of sonar systems.

The generation of sound energy in the water is one function required for an active sonar. The directional and temporal characteristics of acoustic reverberation fields are important for active sonar, while the directional and temporal character of the ambient noise fields are important for both the active echo-ranging sonars and passive listening sonars. The propagation of energy from the source to the target and from the target to the receiver is a phenomenon which directly affects the performance of a sonar system. We are also interested in the extent to which the medium perturbs the propagation and limits the signal-to-noise gains that can be realized with spatial and temporal signal-processing techniques.

A major advance in the technology of acoustic sources which has occurred in the past 10 years is the ability to produce high energy transducers that can generate acoustic power measured in megawatts rather than in tens of kilowatts. This is an achievement which has not come as a flash of inspiration but from a dogged pursuit of materials research and studies in the transmission and generation of power at audio frequencies accompanied by a comprehensive modeling of the interactive radiation and reactive impedances of transducer elements in large arrays (Ref. 1). The development of new piezoelectric materials and the introduction of prestressed ceramic transducers which will tolerate high energy densities are part of this determination to extend the limitations of the fragile piezoceramic materials in the generation of the extremely high pressure stresses required for megawatt transducers. An entire ocean basin can be rung like a bell, putting enough acoustic energy in the water to cause it to reverberate from shore to shore. There is some question as to whether this feat represents a technological advancement or just marks the opening wedge of the sound pollution problem in the ocean. Obviously, if the ocean were to be excited in this manner, the reverberant energy would blind the receiver in proportion to the energy generated.

It is easy to see that more is needed in a sonar system than just high power. Once the level of power is such that reverberation becomes the dominant background, only signal processing gains can bring about an improvement in the system performance (Ref. 2). Signal processing hardware offers a prime example of new ocean technology that should not really be credited to ocean technology. The advances in this area (and there have been many) are founded on the state of the art in digital information processing hardware. This digital technology has been borrowed and exploited extensively (Ref. 3); so much so that it is now reasonable to consider underwater acoustic-array apertures which are comparable in size to those of the large radio-astronomy telescopes.

As is typically the case with improvements, this greater signal processing capability brings us face-to-face with another limitation. Experimental measurements, made with aperture dimensions of several hundred wavelengths, have brought to light the need to be concerned about the stability of the medium. When one attempts to use a large-aperture receiving array to obtain high directional gain, the variability in the sound velocity of the water affects the predictability of the phase of the arrival waveform at any given element of an array and thus limits the beamforming gain which can be achieved with conventional spatial signal processing methods (Ref. 4). A new technology has emerged here also as the persistent effort to crowd the aperture limit has brought us face to face with the problem of working in a nonstationary medium in which significant variations in propagation time occur on both a small and large scale, and has stimulated within the underwater-acoustics community the concept of adaptive and optimum beamforming. This is one new technique which ocean technology can nearly claim as its very own.

These third-generation beamformers are extensions of the electrical phasing networks or time-delay-and-sum beamformers which were used by the German navy in World War II. Because of the large aperture size of some acoustic receiving arrays it is not practical to steer them mechanically as one does the ordinary radar antenna. Thus, for many years electrical phasing networks have been used to generate a directional response from the set of outputs of a hydrophone array (Ref. 5). Whereas this type of electrical phasing is a relative newcomer in the radar game, it is old hat in the sonar game. This is not to say that sonar engineers were any more proficient than radar engineers; it is just that multiple, tapped time delays in the audio-frequency region have been easier to come by and a little easier to handle than multiple time delays in the microwave domain.

Ten years ago the selection of the element configuration for a sonar array was based on a simple isotropic model of the noise field (Ref. 6). One of the reasons for this is that there is a tremendous variability of the background noise level observed in the ocean. The background noise intensity can spread over more than a 40-dB range of level for differing sea conditions and operating areas (Ref. 7). This is akin to the comparison of the noise one would expect in an Iowa cornfield to that one would expect on a busy Chicago street corner. It has, in the past, been hard enough just to build sonars and to attempt to predict their performance with this kind of uncertainty in the model of the background noise. But, this is the present, and, for the past several years, much greater attention has been paid to the directional characteristics of noise (Ref. 8). This concern has stimulated and been stimulated by the new optimum and adaptive beamformer technology.

The optimum beamformers differ from the ordinary delay and sum beamformers in that the actual phase and amplitude weighting functions used for each element are adjusted on the basis of the statistical character of the signal and noise fields to provide for maximum signal response and minimum noise response (Ref. 9). Adaptive beamformers go one step further and optimize in a self-correcting manner for time-varying noise and signal field parameters (Ref. 10). We thus find ourselves at the point of being able now to talk in a realistic way about the use of receiving arrays with high directional gains, gains of 30, 40, and maybe even up to 50 dB in spite of the fluctuation in wavefronts and the anisotropicity of the acoustic fields in which they must operate. Directional gains of this magnitude are so high that it would be essential that the detailed structure of the noise field be taken into account in order to even measure the performance of the arrays. This new technology virtually removes the technical barriers to oceanwide ASW surveillance and enfolds it in the economic constraints that are dictated by the strategic importance assigned to such a capability.

There are other items of technology related to the inner space of the ocean which also impact on our topic. Within the last 10 years a very definite advance has been made in the technology of pressure hulls and floatation materials for deep-submergence use. Titanium is available as a pressure-hull material, at least for small hulls. It offers a significant depth advantage for a neutrally-buoyant pressure hull over even the high strength steels but, unfortunately, at a great increase in cost. Glass and ceramic hulls are still being considered as possible pressure-hull materials but the fabrication technology for these is not yet at hand (Ref. 11). Synthetic-foam floatation materials are commercially available and offer the necessary buoyancy for carrying mechanical systems to great depths. So we can safely say that the depths of the ocean are now open to manned submersibles. Whether or not this greater depth capability has any effect on the arms race remains to be seen.

Other related developments are in the field of vehicle propulsion for submerged bodies. For example, ducted propulsers have shown a marked increase in the thrust efficiency for high-speed bodies and at the same time have reduced cavitation and the consequent radiated noise. Although the past 10 years have not seen much new in the way of nuclear power plants, at least as applied to submersibles, there have been some advances made in the use of more exotic chemical reactions which could be particularly important in looking at the possible applications of high-speed submerged vehicles. These selected items by no means exhaust the new technology in the intra-medium area, but they do highlight those aspects which I think are significant for the topic at hand.

Turning to the surface we find another list of relevant developments. If I were asked to single out the most significant new technology item in the surface category, I would pick the introduction of stable platforms as operating stations in the open sea. During the past 10 years we have witnessed remarkable developments in these platforms, both in commercial rigs for offshore oil exploitation in the open ocean (Ref. 12), and in the oceanography field itself with research platforms such as FLIP, SPAR, POP, and Cousteau's Bouée-Laboratoire, all of these latter being sparbuoy types of stable platforms (Ref. 13). These platforms can well be the precursor of a more extensive technology in extremely large platforms which might find application as logistic seabases for replenishment by aircraft and ship, or perhaps as sensor platforms for radar or sonar surveillance installations out in the open ocean areas.

On the other end of the surface-vehicle spectrum, away from the large stable platforms, we find two vehicle types that have come into vogue in the last 10 years even though their concepts date back, again, many years. One type is the hydrofoil, a high-speed boat which offers moderately high speeds and greater all-weather capability than a rotary-wing aircraft in ASW applications. The other type is the air-cushion vehicle or ground-effect machine which has finally come into its own as a commercially operated vehicle. The GEM may offer a new dimension in logistic support. It can conceivably provide a high-speed transport capability between ship and shore or even ship and ocean platform which not only avoids the requirement for a harbor but can also tolerate more adverse weather in this sea-surface logistic-support role than either fixed- or rotary-wing aircraft.

In the third category, new ocean technology

has crossed over the brink of the abyss. The continental shelf and even the deep sea floor have succumbed to the onslaught of new technology in the past few years. There now exists a technology of deep sea floor work which, by contrast with the capabilities of 10 years ago, could be called brand new. The commercial offshore oil developments augmented by the NSF deep sea drilling project have provided a capability of drilling holes in the deep sea and it now appears as though even hole reentry on the sea floor can be accomplished.

The U.S. Navy's Deep Submergence Systems Project and the small-submersible developments of a large number of private companies have, in the last 10 years, given us the capability of reaching the sea floor with manned submersibles capable of maneuvering near the floor, of carrying out limited manipulation functions for sea floor work, and performing the vital function of inspection and observation in this hostile environment.

A third new sea floor technology is found in the class of remotely operated sea floor, or near sea floor, vehicles and systems. Most of this work to date has been carried out with cable systems because of the great advantages obtained by having an umbilical cord for the transmission of power and control functions. The most striking advances in these areas have been the new cable-towed systems for observation of the sea floor environment and for carrying out geophysical measurements at the bottom of the ocean. Our own vehicle at the Marine Physical Laboratory is typical of others in this general field (Ref. 14). The Marine Physical Laboratory Deep

Tow performs several functions. It generates high resolution side-looking sonar pictures of the sea floor for both bottom search and geophysical survey work. By use of a narrow-beam echo sounder it provides information on the fine-scale topography of the sea floor. A very precise acoustic-transponder navigation network allows documented bottom photographs to be correlated with both the fine-scale topographic maps obtained with the Deep Tow system and with the corresponding acoustic structure of the bottom as observed with the side-looking sonar. These developments have brought multisensor observation of the deep sea floor within the state of the art.

Not quite up to the same stage of development we find the remotely operated sea floor work systems which extend beyond the observation capability of the towed sea floor search vehicles by providing a means of modifying the sea floor environment. These remotely operated sea floor work systems are strong economic competitors to manned submersibles for the work tasks associated with the installation and for recovery of sea floor structures and equipment (Ref. 15).

These are the highlights of our new ocean technology which expand our horizons in viewing the arms race. The overall impact of these new technologies on the arms race is probably somewhat neutral. They can, on the one hand, be a factor in competitive arms escalation, particularly in the strategic area, or on the other hand, they can be used in an effective way for inspection and monitoring in the control of the de-escalation of armament.

The combination of new acoustic technology

and that of the stable surface platforms gives rise to a new freedom in ocean surveillance. Horizons are expanded in this area to a technically unlimited capability of monitoring the oceans with a finite number of sonar systems for the locations of submarines which are submerged below the reach of land, ship, airborne, or satellite electromagnetic-surveillance systems.

The new family of surface platforms makes it possible to create large pieces of "real estate" in strategic locations in the open ocean where they are desired and when they are required. Logistic support for these bases can be provided with the new surface craft. No longer must operations be restricted to those carried out either from bases in a foreign nation's national territory or from mobile ship platforms at sea.

Capabilities in both anti- and pro-submarine warfare have been extended by virtue of the increased operating depths for submersibles and the capabilities of the new classes of hydrofoil and ground-effect surface vehicles.

The bottom of the sea is accessible for installations of a variety of types. Man could get there, could work there, and could, in effect, remain there for indefinite periods of time if he were to make use of the remote control and sensor techniques which are in hand today.

We should realistically consider the impact of strategic bases in the open ocean away from the continental land masses—not only the large stable surface platforms, which are relatively easy to support logistically but perhaps somewhat vulnerable to attack, but also bases located below the surface as sea floor installations (Ref. 16). A sea floor installation would obviously be far more difficult to support logistically but could conceivably be of a covert nature. The concealment offered by the ocean column should not be dismissed lightly; it can be very effective.

The very cursory and quite general nature of this discussion has been dictated by the complexity of the ocean technology. As a change of diet it might be interesting to touch upon one specific application in more detail.

Consider, as an example, the possible advantages of sea-based seismic arrays for monitoring nuclear tests (Ref. 17). At present the most promising discriminant used for distinguishing nuclear explosions from earthquakes is the spectral difference between the very low frequency region—0.01 Hz—and the spectrum in the vicinity of 10 Hz. At the present time the large land-based seismic arrays are noise limited in the low frequency range, 0.01 Hz to 0.1 Hz, and this noise limit seems to be meteorologically induced. If a seismic installation were made on the deep sea floor, the seismometer would be of the order of 4 wavelengths below the surface where the meteorological noise originates, and a considerable attenuation of the low frequency noise would be expected. This has not yet been confirmed, but experiments for this measurement are being planned.

A major advantage of a sea floor array is that there are vast provinces at desirable locations in the oceans where installations could be made over a very regular mantle structure, which should provide better signal wavefront coherence than that experienced in the continental array installations. Thus larger arrays could be utilized effectively.

If the lowest noise were to be achieved, the deep drilling technology would have to be used to implant the seismic sensors several hundred feet deep in the sea floor to eliminate the very low frequency background noise associated with the flow noise that is generated by even the slow (0.1 knot) deep-sea currents. Why is this important? If this lower noise level can be realized, the seismic detection capability threshold could be lowered down to a fraction of a kiloton. There is real doubt if any useful weapon development could be carried out at these very low yields. Thus a full monitoring capability could exist for enforcing a total test ban.

I would conclude by emphasizing that the new ocean technology will have a definite impact on the arms race—be it an escalation or de-escalation phase in which we find ourselves in the future.

References

1. C. H. Sherman, "Analysis of Acoustic Interactions in Transducer Arrays," *IEEE Trans. on Sonics and Ultrasonics,* vol. Su–13, (March 1966), pp. 9–15.

2. J. L. Stewart and E. C. Westerfield, "A Theory of Active Sonar Detection," *Proc. IRE,* vol. 47, no. 5 (May 1959), pp. 872–881.

3. W. B. Allen and E. C. Westerfield, "Digital Compressed Time Correlators and Matched Filters for Active Sonar," *J. Acoust. Soc. Am.,* vol. 36, no. 1 (January 1964), pp. 121–139.

4. J. C. Beckerle, J. L. Wagar, and R. D. Worley, "Underwater Acoustic Wavefront Variations and Internal Waves," *J. Acoust. Soc. Am.,* vol. 44, no. 1 (July 1968), p. 295.

5. J. W. Horton, *Fundamentals of Sonar,* U.S. Naval Institute, Washington, D.C., 1959, chapter 5.

6. V. C. Anderson, "Arrays for the Investigation of Background Noise in the Ocean," *J. Acoust. Soc. Am.,* vol. 30, no. 5 (May 1958), pp. 470–477.

7. R. J. Urick, *Principles of Underwater Sound for Engineers,* McGraw-Hill, New York, 1967, chapter 5.

8. E. H. Axlerod, B. A. Schooner, and W. A. Von Winkle, "Vertical Directionality of Ambient Noise in the Deep Ocean at a Site near Bermuda," *J. Acoust. Soc. Am.*, vol. 37, no. 1 (January 1965), pp. 77–83.

9. F. Bryn, "Optimum Signal Processing of Three Dimensional Arrays Operating on Gaussian Signals and Noise," *J. Acoust Soc. Am.*, vol. 34, no. 3 (March 1962), pp. 289–297.

10. D. J. Edelblute, J. M. Fiske, and G. L. Kinnison, "Criteria for Optimum-Signal-Detection Theory for Arrays," *J. Acoust. Soc. Am.*, vol. 41, no. 1 (January 1967), pp. 199–205.

11. *Oceanology International*, March 1970 (special issue on marine materials).

12. "Mobile Rigs Completed or Under Construction This Year," *Ocean Industry*, vol. 4, no. 11 (November 1969), pp. 76–78.

13. F. N. Spiess, *Ocean Engineering*, J. F. Brahtz, ed., John Wiley & Sons, New York, 1968, chapter 15.

14. F. N. Spiess and J. D. Mudie, *The Seas*, vol. IV, A. E. Maxwell, ed., John Wiley & Sons, New York, 1970, chapter 7.

15. V. C. Anderson, "Maintenance of Sea Floor Electronics," *IEEE Trans. on Aerospace and Electronics*, vol. Aes–4, no. 5 (September 1968), pp. 650–658.

16. W. B. McLean and J. Newbauer, "A Bedrock View of Ocean Engineering," *Astronautics and Aeronautics*, April 1969, pp. 30–36.

17. E. C. Bullard, "The Detection of Underground Explosions," *Scientific American*, vol. 215, no. 1 (July 1966), p. 19.

Technical Means for the Investigation and Exploitation of the Ocean

I. E. Mikhaltsev

Various technical means for the study and exploitation of marine resources have been developed during the last years. The prospective growth of food and mineral resources, oil and gas obtained from the ocean, as well as the requirements of global weather forecasts and improvement of navigation, call for new technical means to work in the ocean.

The validity of our assertion becomes evident even from the briefest consideration of the situation.

For instance, the effective finding and catching of fish and sea animals require the use of acoustic multifrequency echo-locators and high-speed midwater trawls. Exploration and mining of mineral resources—prospecting for iron-manganese nodules, heavy mineral deposits, recovery of chemical materials from seawater and, at last, oil and gas prospecting on the shelf and in the deep ocean—make necessary the use of new and improved means of experimental geophysics, mining engineering and other special technical means. Shipboard and bottom gravity meters, towed magnetometers, seismoacoustic towed systems and narrow-beam echo-locating devices for continuous profiling of the ocean bottom form the basis of the marine geophysical complex. A large group of the most complicated technical procedures is used for oil and gas recovery from the ocean bottom. This group includes floating and stationary drilling equipment, underwater television apparatus,

Igor Mikhaltsev is Deputy Director of the Institute of Oceanology of the Academy of Sciences of the U.S.S.R.

acoustic bottom beacons for a precise determination of the place of floating equipment and for keeping it at that point, and in addition, various deep-diving equipment enabling man to work at the ocean bottom. The list of the instruments would be incomplete if no mention were made of bottom dredges and special acoustic means for the near-bottom survey.

Weather forecasts on our planet, two-thirds of which is covered by the oceans, require knowledge of the physical properties of the oceanic water; in particular, of its temperature structure and variability of the ocean currents, as well as the parameters of the ocean-atmosphere interaction. The land network of meteorological stations providing the necessary information should have an ocean equivalent in the form of a global complex of anchored buoy stations supplied with telemetering systems for data transmission through artificial earth satellites. It should be noted that the ever-growing use of the oceans for transport purposes makes the knowledge of the ocean currents, waves, and water temperature as indispensable here as it is in other fields. The use of satellites and acoustic aids has resulted in the improvement of navigation techniques which are successfully used in addition to radar instruments. The prospective use of underwater cargo vehicles, and specifically of underwater tankers, may enhance the importance of the acoustic methods of underwater studies and navigation.

Technical means enabling man to penetrate to the ocean depths are, to a certain extent, of a peculiar significance. They include, first of all, hermetic self-propelled vehicles, i.e., deep-sea, manned submersibles used mostly for research or rescue operations. A great variety of such underwater vehicles that have been developed in recent years and that are meant for submersion to depths of several thousand meters are based on the achievements of metallurgy, automation and hydroacoustics. The second trend in the development of diving techniques is the improvement of technical means to give man the possibility of free movement, life, and work in the water. The latter means include the aqualung working with different gas mixtures, underwater laboratories for a long stay of aquanauts at depths, and the improvement of high-pressure and decompression chambers.

Among the technical means related to the exploitation of the seas and oceans there are such specific technical facilities as those for rest and sports in the sea (for example, scuba equipment, light boats, etc.), as well as technical means for determining the degree of water pollution and averting it.

It can easily be shown that all the above-mentioned technical means and instruments can be of a direct or indirect military importance. The specific features of the ocean can make it an object of studies for scientists, the international scene for a wide cooperation, or the theater of war. The possibility of using the ocean for military purposes may be made smaller if it is studied cooperatively by various nations and if large technical systems and devices for its investigation are exploited internationally. Cooperative studies can help us to eliminate both the problem of keeping secret results obtained in the geophysical investigations and the possibility that they will be used for military purposes. The Antarctic investigations are good examples of such fruitful cooperation. Some more specific examples of such a perspective for the ocean can be mentioned. The creation and exploitation, on a wide international basis, of a global system of oceanic buoy stations is an important possibility. This work has been initiated by UNESCO. However, there are many obstacles in the way of its progress. An international program of man's penetration to the ocean depths may be another possibility. A comprehensive program of cooperative studies aimed at providing the necessary technical means and finding the optimum physiological conditions for man in the sea can be based on the well-known nonmilitary projects now in progress in France, the United Kingdom, the United States, and the USSR (such as "Conshelf," "Tektite," "Chernomor," etc.). The same refers to the possibility of the program for deep-sea studies from research submersibles.

Finally, the example of the International Geophysical Year shows the possibility and efficiency of a wide many-year international program for studies of the whole World Ocean.

I think that the spirit of Pugwash makes us think of countermeasures to the arms race and the military use of new achievements in the knowledge of our planet. A close cooperation among the scientists of various specialities can make the ocean a scene of cooperation in the use of new technical means for the exploitation of marine resources for the benefit of mankind.

Summary of Discussion of the Papers on Ocean Technology

Nonacoustical Detection of Submarines. There are several methods available which do not depend on sound for detecting submarines. Differential infrared radiometers in satellites may be used to measure surface temperature differences of the order of hundredths of a degree (provided the temperature differences extend over sufficiently large areas). This is a powerful tool for studying ocean currents. Although large amounts of energy are required to create these temperature differences, their general significance and their relationship to submarines are both unknown. If a submarine is near the surface and churns up the water, just this turbulence will give rise to temperature differentials and differences in reflectivity that can be observed. However, similar turbulence occurs from natural currents in the ocean.

Electromagnetic radiation is very limited in its applicability in the ocean. There is but one window, which occurs in the green portion of the visible spectrum. The only advantage of using lasers is that they provide a collimated beam, which can have some marginal advantage in terms of scattering. But absorption is very strong and even if the penetration distance could be increased by an order of magnitude, it would still be very short. The detection of submarines through the surface disturbances that they create is difficult because of the background due to wind-driven wave patterns. Magnetic detection is governed by a third-power law and so suffers greatly from depth limitation. In summary, nonacoustical methods of detection offer little promise for the future when compared with the advantages of acoustical methods.

Active and Passive Sonar. If one looks at the general problem of acoustically detecting a submarine in the ocean, there is no clear-cut choice between employing active sonar (sending out a signal and then listening for a reflection) and passive sonar (simply listening for noise generated by the submarine). The operational constraints call for both capabilities. This is so because the type of system employed by, or anticipated for, the antisubmarine forces influences both the design of the submarine and its operating mode. The countermeasure to a passive sonar is to quiet the submarine and run slow, keeping machinery noise, turbulence, and flow noise to a minimum. It is easy to drop the radiated noise from a submarine by 20 or 30 dB, possibly taking it out of the range of the passive sonar. This mode of operation requires that a submarine remain in an area for a very long time. On the other hand, the use of active sonar alerts the target that he is within or near the range of your surveillance. To escape, he may then shift into a high-speed operating mode so that the surveillance system does not know where he *is*, but only where he *was*. With acoustic propagation times of minutes this delay can be a significant factor.

Accuracy of Sonar Detection. The relative accuracy of detection is not particularly limited by range. In general, one could count on an accuracy of one-tenth of one percent of the range, i.e., one-tenth of a mile accuracy at one hundred miles. Greater accuracies are not needed, since a surveillance system would not be expected to have fire-control accuracy. Rather, the indication of the presence of a submarine would be sufficient to call in a secondary search system. This latter system would have a relatively small area to search and could pinpoint the target.

Hostility of the Ocean Medium to Sonar. The ocean medium, especially because of its refractive characteristics, is rather hostile to sonar. In general, there will be some areas where the submarine will be invisible and other areas where it will not. The location of these regions is hard to predict, because they depend on the exact thermal structure of the water. The phase reversal of reflected sound at the surface boundary poses another difficulty. Because of this effect the submarine seeks to travel near the surface in an effort to evade active sonar. In general, the submarine is more detectable if it is deep. However, near the surface it is more susceptible to magnetic detection techniques, or even radar, if it breaks the surface. Reverberation from the bottom and the classification of the detected object (to distinguish a submarine from a school of fish) also cause real problems for the active sonar system. Doppler discrimination can be used to help alleviate these difficulties.

Detection on the Continental Shelf. The problem of detecting submarines is much more difficult on or over the continental shelf than in the open ocean because of multiple reflections from the bottom. These cause loss of coherence in the wave front and make the signal processing much more difficult. The surveillance potential predicted in Dr. Anderson's paper really applies to the open ocean only.

Cost and Structure of an Ocean-wide Sonar Detection System. The amount of money required to provide ocean-wide ASW capability was estimated to be in the billion-dollar category, but not tens of billions, assuming no effective countermeasures were taken. The numbers and positioning of both the transmitters and the detectors depend on the range of the system, which in turn depends very much on the water depth and the particular transmission characteristics of a region of the ocean. In the Pacific, which is very deep, sound can travel as a purely refractive wave, never hitting the bottom, resulting in an effective range of hundreds of miles. In this case the delay time would be long and this is operationally very significant. In other areas the range may be only a few miles. If the sound propagation follows an inverse-square law or less, the best configuration for receivers is to cluster them together and to phase them into a directional array.

Impossibility of Building a Clandestine Ocean-wide ASW System. No one could build an ocean-wide ASW system without it being very visible and the other side being very aware of it. Hence the worst-case analyst need not worry about a surprise in this field. Furthermore, the large size and extent of the system tends to make feasible outlawing it by an international treaty. It would not be possible to build surreptitiously an ocean-wide sonar system.

The Possibility of Countermeasures. Even if such a system were built several alternatives would be open to the other side. In the first place a surveillance system could never be 100% sure. There would always be places where a submarine could hide because propagation is chopped up. But the system could be much less than 100% sure and still render ineffective a sea-based deterrent system. On the other hand, the response to a full-ocean acoustical surveillance system could be to ring or to jam the entire ocean with another large system of buoys. If the jammers were only fixed point sources, adaptive processing techniques would be highly effective against them. But if the jamming were done by ships moving in the area, the discrimination problem would be much more difficult.

The Legal Regime of the High Seas. The legal regime of the high seas is very unclear today in the face of these new technologies. The question of sovereignty over large ocean platforms has not been settled, nor has the degree to which the presence of such a platform would be allowed to restrict free passage of other vehicles in its immediate vicinity.

Implications for the Undersea-based Deterrent. Following the discussion on ocean technology, Dr. Kistiakowsky said for the record that he did not believe that the new possibilities in sonar really jeopardized the undersea-based deterrent. To think that they did, he indicated, would be to make the mistake that was alluded to earlier, namely "the fallacy of the last move." There are in fact countermeasures possible. The Chairman expressed agreement with Dr. Kistiakowsky.

Summary of Discussion on New Technology Based on Comments by R. Sagdeev

It is useful to summarize briefly the results of our technical investigations in these few days (see Table 1). The table has a double system of classification: under area of application and area of technology. Let us start with nuclear physics. There was wide agreement with Dr. Mark's comment that, in the sense of creating new weapons which could change the strategic balance, nuclear physics is now exhausted. However the possibility of producing plutonium weapons from reactor-grade material increases the ease of the spread of nuclear weapons. The superheavy-element weapon was reported to be quite unfeasible. Anyway, it would not change the strategic balance and probably would have only tactical importance. The possibility of triggering a D-T mixture, using either a focused laser or a focused electron beam, is remote with today's technology and, again, even if accomplished, this would not change the strategic balance. On the other hand, a high-altitude nuclear explosion, resulting in a long lasting plasma cloud which could black out an ABM radar, may have some effect on the overall strategic balance. However, because of the test ban treaty, there is insufficient experimental data available. But perhaps this plays a useful role in discouraging the deployment of ABM systems.

Magnetohydrodynamics may provide new opportunities for jamming radar. This again would not change the strategic balance and may help to dis-

Table 1 Summary of New Technology and Its Effect on the Arms Race.

Area of Technology	Area of Application			
	Strategic Arms		Reconnaissance and Surveillance	Tactical
	Offense	Defense		
Nuclear Physics	Plutonium weapons from reactor-grade material; high altitude explosion blackout			SHE-weapons
Lasers		Laser gun		D-T trigger
Magnetohydrodynamics	Jamming radar			
Plasma	High altitude explosion blackout			D-T trigger
Sensors; Infrared, Radio, Other Electromagnetic Radiation	Jamming	Mid-course ABM, down-looking radar, terminal ABM, radiation gun	Transparent boxes, infrared and ultraviolet optical reconnaissance	Down-looking radar
Ocean Technology	Ocean floor or continental shelf installations		Sonar on surface platforms for ASW	
Guidance	Improve accuracy			
Data Processing		ABM systems		

courage ABM deployment. Laser technology also provides the possibility of a laser gun. The first experiments with such a weapon described in the open literature were offensive weapons in which laser cowboys tried to shoot cows. But such a weapon has probably more potential in a defensive role; in principle, however, there is also the possibility of a radiation gun playing an offensive role.

As far as radio, infrared, and other electromagnetic radiations are concerned, these provide opportunities for jamming many sorts of sensors, as well as providing the working medium for the transparent boxes spoken of by Dr. Fubini. There was, however, scepticism about the usefulness of these boxes. The possibilities were also mentioned of airborne, down-looking radar and of improvements in terminal ABM radar. Even a small flyby sensor to aid a mid-course ABM system was suggested.

There appear to be several new prospects in the future arising out of ocean technology. In the future lies the possibility of offensive installations on the continental shelf or even the deep ocean floor. In the long run such installations could increase stability. Both Dr. Anderson and Dr. Mikhaltzev described surface platforms or ocean buoys for submarine-detecting sonar arrays. Although Dr. Anderson was optimistic that the ocean could be made transparent, it was pointed out in the discussion that a large number of sonar platforms would be needed and that there are possibilities of jamming such systems. So the idea of making the ocean transparent may not be realistic today.

Quite surprising was Dr. Hoag's prediction that

guidance technology would improve the accuracy of ICBMs to 30 meters in ten years. However, to consider the strategic consequences of this prediction, one must remember that this accuracy is relevant only for fixed land-based targets.

Progress in data processing techniques will contribute to the development of an ABM system.

The scope of the symposium was dictated to a large extent by the expertise of the people who attended and who presented papers. Therefore the review of technology was surely not exhaustive. For example, nothing was said about the possibility of using weather modification in a belligerent way, or of developments in chemical and biological warfare.

The opinion was stated that, from a technological and scientific point of view, nothing is coming in the next ten or so years which would completely change the strategic balance. Therefore there is no reason to exhibit fear or accelerate military research and development.

On the other hand, it was pointed out that there is a large and influential group of people who would consider some of the projections and predictions made at this conference very destabilizing. The prediction of 30-meter accuracy for ICBMs and the suggestion that the technology may be at hand to make the ocean transparent may cause great worry to some people. Since these people are influential, the result may be to spend a great deal of money to counteract or compensate for those possibilities. Those who do not share those fears have a major educational role to play if they wish to prevent a new escalation of the arms race.

2 Safeguarding Nuclear Installations

Introduction to Part Two

At the end of World War II, the technical knowledge necessary to harness the energy of nuclear fission for productive purposes was restricted to very few countries. In the ensuing years, however, this knowledge has spread around the world and is being ever more widely exploited. Whatever the economic benefits and environmental risks of using nuclear energy as a commercial power source, there is one danger that has been universally recognized from the beginning: the possibility that the spread of peacetime nuclear technology will be accompanied by the spread of nuclear weapons.

A major by-product of power reactors containing uranium is the element plutonium. If the fuel rods are removed from a reactor after a relatively small amount of the uranium has been consumed, the plutonium will consist almost entirely of the isotope Pu^{239}, an isotope from which nuclear weapons can be conveniently made. With longer exposure increasing amounts of Pu^{240} as well as Pu^{239} will be formed. Some years ago, it was widely believed that plutonium with an appreciable concentration of the heavier isotope would not be suitable for weapons use. This seemed to provide a hope that, despite the anticipated widespread production of vast quantities of plutonium in a rapidly expanding worldwide nuclear power economy, the reactor fuel cycle could be designed so as to produce plutonium of this "denatured" variety; if this were possible, the problem of diversion of plutonium from peaceful to weapons use could have been eliminated or at least greatly reduced. Unhappily, this hope has proved to be illusory; the degree of Pu^{240} concentration attainable in reactors of feasible design does not appear to present a serious obstacle to weapons construction.*

Another bar to the easy proliferation of nuclear weapons has been the technological difficulties and expense involved in the separation of the readily fissionable U^{235} from natural uranium, in which it is present in very low abundance (1 part in 140). Here, again, advancing technology has frustrated arms controllers; ultracentrifuge technology has made possible separation plants that are both considerably smaller and more easily constructed than the diffusion plants heretofore required for the economical concentration of the U^{235}. Although nuclear-power requirements demand considerably lower concentrations of U^{235} (in the 1% to 10% range) than are needed for weapons production (80% or more), once plants capable of producing the low-enrichment material have become available, relatively less effort and material will be required for the ultracentrifuge facilities to raise the concentration of U^{235} to the weapon-grade level.

These advances in technology, coupled with the anticipated explosive increase over the next decade in world-wide power consumption and the increasing attraction of nuclear-power reactors to satisfy the demands for cheap power, serve to exacerbate the real dangers of diversion of the materials and by-products of commercial atomic

* See the paper on "Nuclear Weapons Technology" by J. C. Mark.

power generation to the production of nuclear weapons.

Along with the development of nuclear power technology, the technologically advanced nations have developed various means of preventing the diversion of nuclear materials into the military realm. These efforts have centered on so-called safeguard procedures—methods of control and inspection of nuclear facilities to assure against illicit diversion of militarily useful materials. Most of these methods have been applied on a bilateral basis, through agreements between the technically advanced suppliers and the less-developed recipients of their aid. The coming into force of the nonproliferation treaty has highlighted the importance of universally accepted safeguard procedures. The nonnuclear signatories of this treaty have accepted the principle that all their commercial nuclear installations will be safeguarded. Although the treaty is now operative, not all the technologically advanced nations have yet signed it, nor have the procedures been yet worked out to fulfill it.

It had not been the original intention of the organizers of this symposium to include the subject of safeguards on the agenda. However we soon became aware of considerable expertise among the participants in the symposium and a wide interest in the subject. As a consequence, some time was set aside for a discussion of the safeguards problems and, although no formal papers had been solicited beforehand, a discussion developed almost spontaneously. This discussion was highlighted by several lengthy presentations and a number of briefer interventions. All of these, as well as a summary of the general discussion, are included here in the hope that they may contribute to the continuing search for agreement on this important aspect of arms control.

Safeguards of Nuclear Materials

W. Häfele

The problem of safeguards is fairly old. It has been a concern since 1945 or 1946, since the time of the Baruch Plan. However, safeguards have really come into focus only recently for quite a number of reasons. It is only recently that the use of nuclear energy has become worldwide and on a large scale. It is only recently that the necessities for safeguards are really that large, as only recently has the use of nuclear energy become of commercial significance. This necessarily then brings into the picture large amounts of U^{235} and plutonium. These large amounts are handled in the civil domain of nuclear energy as distinct from the domain of military applications. It has been recognized more recently that it is the nuclear material which possibly interlinks this civilian domain to the military domain, and this possible interaction has raised concern. Among other things, it led to the nonproliferation treaty, which in turn raised the point of building a modern safeguards system. For two or three years now there has been an international debate on what such modern safeguards could look like.

I think there are three major points that characterize such a modern safeguards system. These three points are: a modern safeguards system must be rational, objective, and formalized. It must be rational because of the overwhelmingly large size of the safeguards job. It will be internationally and generally acceptable only if the cost burden is some-

Dr. W. Häfele is Scientific Director of the Institute for Applied Reactor Physics in Karlsruhe, F.G.R.

how reasonable and not overwhelming; therefore it must be rational. At present at the International Atomic Energy Agency in Vienna this question of safeguards costs is thoroughly debated among other points. We may be tempted to say that this question is irrelevant, but as long as we live in a world of economic progress and economic equilibrium, the cost considerations for safeguards is to my mind relevant. The second point is that modern safeguards have to be objective. If safeguards are to succeed on a worldwide basis, they must be acceptable to both the inspector and the inspected. The subjective feeling of one of these two is an insufficient basis for a successful safeguards scheme. If a modern safeguards system is to be worldwide and unique and to embrace all nations or groups of nations, potential conflicts would be a built-in feature of the system. Conflict could arise, for example, from either unfounded accusation and excessive thoroughness on the inspectors' part, or a too-relaxed attitude on the side of the inspected, perhaps in order to be nationally competitive, without either side having in mind a violation of the treaty. If by objective safeguards the terms of communication between inspector and inspected are predetermined, it helps to make the whole safeguards job manageable. The third point is that modern safeguards have to be formalized. Again, formalization of the inspection scheme helps the inspector and the inspected. There is an additional feature which requires formalization, and this is the inherent open-endedness of most of the inspection procedures. Take for instance the case of a fabrication plant or, if you prefer, a chemical reprocessing plant. Given a throughput of say 100 kg or a ton per year, do you search for the last 100 grams, for the last ten grams, for the last gram or even further? Where is the limit? If you handle nuclear material openly and you want to close the material balance, you always face that inherent open-endedness, because there are always natural losses, that is, material unaccounted for. Therefore it is necessary to preestablish certain limits beyond which the inspection is satisfactory and below which additional action is required. Together with the necessity of having an objective system, these two points lead also to the question of having formalized statements by the inspector and possibly also by the inspected. These formalized statements must be based on a statistical or operational research approach, so that ultimately only certain figures, characterizing the material unaccounted for or the confidence level, would have to be inserted. One can think of an analogue to all this; for instance, I think traffic control is such an analogue. Traffic control is highly formalized. At green everybody is allowed to go and at red everybody has to stop. The green and the red lights don't make any distinction between any participants in the traffic. It is predetermined, which means that these measures concentrate only on letting people go or not go and nothing else. The more general concept behind that point is that of limited mutual trust. Full mutual trust cannot be the basis, because otherwise the safeguards would not be necessary at all, but it is also impossible to entirely exclude mutual trust; therefore, the point is that there must be limited mutual trust and therefore objective and formalized cooperation between the inspector and the inspected. Bearing in mind these three points of requiring a modern safeguards system to be rational, objective, and formalized, one can ask what are the principal components of such a modern safeguards system? How can it be built? There is now general agreement that there are principally three measures, namely "containment," accountability, and surveillance.

"Containment" comes into the picture because it is the nuclear material that establishes the link between the civil domain and the military domain. Therefore one has to ask for measures that make it impossible to transfer this nuclear material from the civil to the military domain. In major nuclear facilities, such as reactors, reprocessing plants, or fabrication plants, there are features of containment. These are, by their very nature, a help for safeguards. The containment of a nuclear reactor inhibits a large number of unauthorized fuel shipments, and there are only particular doors where fresh fuel or burnt fuel can come into the reactor or leave it. In the case of a reprocessing plant, there are only a few entrances where irradiated material can readily enter the plant. Also it will be hard, in the case of a reactor, to bring unauthorized fuel into the reactor for diverting neutrons, if the containment is a well-established feature of that reactor. Of course it is necessary to extrapolate the containment function. In the context of safeguards, safekeeping and sealing of transportation casks are also a safeguards measure. Personnel locks, going in or going out of a certain area, are also part of that containment measure.

The next safeguard measure is accountability, and this may be the most important one. Accountability means mass balance. You take into account the inventory, the inputs, and the outputs of nuclear material with respect to a given nuclear facility. It is the most important safeguards measure, because it produces figures, and figures are best suited to support the point of objectivity and formalization. In many cases it is possible to have control over a large part of the fuel cycle just by counting pieces. One can call it digital accountability, that is, to count the number of fuel elements from the fabrication plant going through the reactor and going into the reprocessing plant. There is the other part of the fuel cycle—where the material is openly handled, from the reprocessing plant into the fabrication plant—and therefore one may call this the part of the fuel cycle which has open accountability. Operational losses and errors explain the material unaccounted for and one has to look in detail at what the possibilities for establishing the mass balance really are.

The third measure is surveillance. This means visual inspection by human beings. This scheme is in certain cases the easiest, although it does not lead to an objective situation. In all cases of surveillance the ultimate result is a subjective conclusion. Therefore, if one has a choice between the three components for building up a modern safeguards system, namely, containment, accountability, and surveillance, undoubtedly one would opt for accountability wherever it is possible to do so. If one looks in greater detail into the matter, one comes to the point of asking in each particular case for the optimum combination between these three principal components. In so doing, one has to realize that it is practically impossible in most cases to rely on only one measure alone and that one has always to ask for a certain degree of redundancy. Therefore, probably, mass balance always goes together with some containment measure and maybe some measure of surveillance. Another important point is that the whole inspection process is in itself open-ended. This means that it is practically impossible to make 100% sure that absolutely no fissionable material has been diverted over an extended period of time. If you are handling, for instance, tons of plutonium in the civilian domain and if you are ready to wait sufficiently long, say ten years, you always can construct cases where, say 10 g or so of plutonium could have been diverted without your realizing it. Therefore one comes to the conclusion that the objective of modern safeguards is only to reduce very strongly the probability of diversion, but not to inhibit it with 100% confidence. It seems to me to be very important that both the political side and the technical side understand each other with respect to this point.

Now let me come to the next step: If you employ accountability and containment as two principal safeguards measures, then you are automatically led to concentrate these safeguards activities at a number of points. In a properly designed containment the entrances and the exits will be accountability points of very high importance, key points or, as they have been called in the text of the nonproliferation treaty, strategic points. It is possible to concentrate these safeguards activities there and to leave the material balance area between two strategic points, say, between a point of entrance and a point of exit, largely untouched. Therefore such an approach helps to make modern safeguards, to a very large extent, unintrusive. And let me come to another point: modern safeguards are safeguarding the whole fuel cycle and not only the isolated nuclear facility as has very often been the case in the past. It is a typical feature of the post-nonproliferation era that it is such a whole fuel cycle that is to be safeguarded. This gives a number of operational improvements, as there are quite a few interdependences in the fuel cycle. For instance, when a batch of fuel elements for a light-water reactor leaves the fabrication plant, its uranium content is known to better than 1%, both the absolute uranium content and the U^{235} content. If you make sure that this fuel passes the reactor and really arrives, after its residence time in the reactor, at the reprocessing plant, you can accurately determine the remaining uranium content. However, it has always been operationally difficult to determine the uranium input into the reprocessing plant by purely chemical means. According to some detailed investigations it seems to be possible, therefore, to ask at the entrance of the reprocessing plant, for safeguards purposes, only for ratios and no longer for absolute measurements.

Taking into account these interdependences and the concentration of safeguards activities at certain points, it is only natural now to ask for the development of a number of instruments for measuring the flow of fissionable materials into and out of these principal nuclear facilities. In developing these instruments one can distinguish between direct methods and indirect methods: direct methods for the part of the fuel cycle where the material is openly accessible, and indirect methods for that part of the fuel cycle where the fuel is contained; the indirect methods are largely for nondestructive testing. The main point of using indirect methods is the existence of the fabrication plant. Calorimetry is employed there in order to determine the plutonium content of a fuel. Provided that the isotopic composition of the plutonium is known, accuracies of better than 1% can be achieved. In case of the uranium fuel for light-water reactors, the lead-pile spectrometer is being employed to determine the U^{235} content to the order of 3% or 4%. An example of a direct method is the x-ray fluorescence which is used to determine the plutonium and uranium content in the presence of highly radioactive fission products and therefore is most suitable for the entrance of the reprocessing plant. An automated mass-spectrometer station can automatically handle samples taken from the process line of the reprocessing plant. Recently also the technique of delayed neutrons has been employed for both nondestructive testing and determination of the content of waste streams. I will not go into these methods here in greater detail. It is a very fascinating field, but it is straightforward physics and chemistry, and its methodology is well known. Rather, I will emphasize the more unusual aspects of safeguards, namely, putting everything together into a system.

For this, it is necessary to pursue a thorough systems analysis. Systems analysis, in this sense, is the study and quantitative interrelating of the interactions of the various components of the system, including human beings. It is necessary to have certain yardsticks. One most important yardstick has been defined already some years ago, namely that of effective kilograms. Effective kilograms is the figure that comes out if the actual kilograms of nuclear fuel is multiplied by the square of the enrichment of that material. It refers to the concept of criticality and distinguishes between low-enriched material and high-enriched material with respect to their safeguards value. Another yardstick that only recently came up is that of critical time. That means that the safeguards systems must be designed in such a way that in closing the mass balance, the frequency of inventory takings or similar aspects are determined in such a way that they are required only at certain time intervals. This time interval is determined by the time necessary to divert material into the military domain, making reference to the initial conditions. If the material is of low enrichment and in oxide form, a large critical time is possible; but if it is highly enriched and in metal form, only a short critical time is allowed. In pursuance of the systems analysis the cost-benefit aspect also comes into the picture. Various approaches are possible and thereby also various distributions of safeguards efforts in the different parts of the fuel cycle. In the past, the reactor has been the focus of interest. I think this has been the case because the reactor both breeds plutonium and is the most frequent and most interesting facility. But nowadays the concentration is more on the reprocessing plant and the fabrication plant, where the material is openly handled. These differ from reactors, where the material is contained in sub-assemblies. It is easy, as I said before, to count the fuel elements and to make sure that they ultimately arrive at the reprocessing plant. Therefore it may very well be that the safeguards efforts for reactors—and that comprises 80% of all the cases—may be fairly limited and ultimately easy. So a project for building up modern safeguards has, altogether, at least three major areas: the development of instruments for indirect methods, the development of instruments for direct methods, and the related systems analysis.

Finally I will make a remark about the methodology of building up a modern safeguards system. In normal science, as we are all familiar with it, the object of the investigation is nature. The scheme of investigating nature is principally objective and the methods are therefore invariant with respect to time and to space. In other words, to make an experiment independent of the persons executing it is the principal condition in pursuing science. The main tools for science are mathematical models and these mathematical models necessarily do not include human behavior. Making a mathematical model of a system, and specifically of safeguards, is in principal a different matter. True, the behavior of nature—for instance, measuring the mass flow at certain points—also comes into the picture here. But the action, interaction and possible conflict of human beings, the inspector or the inspected, also must ultimately be described by a mathematical model. Typically, this has led a number of my colleagues at the University of Karlsruhe to propose the employment of the theory of games, which indeed is specifically designed to express a conflict situation. It may be possible by such an approach to quantify the aspect of surveillance as a safeguard measure and to balance it against accountability and containment, thereby also evaluating the desired degree of redundancy. Of course, in order to do so, one has to define payoff functions. We did this, in the beginning, to identify sensitive parameters as against nonsensitive parameters. But now it has emerged more and more that only certain inequalities are necessary in the regime of payoff functions in order to arrive at some essential and already important conclusions. Maybe other mathematical tools can also help finally to build up a complete mathematical model of such a safeguards procedure. So the methodology of investigating safeguards turns out to be very fascinating and challenging, contrary to what we believed in the very beginning, when we first attacked this problem.

I believe that developing the methodology of such a large worldwide and unique safeguards approach is a very important thing, because it is almost certain that in the next ten years it is not only nuclear material that has to be safeguarded but also many other things like air, oil, water, etc. Therefore I feel that the initial challenge of doing the safeguards job right has to be met in order to be able to meet the requirements of the forthcoming ten or twenty years.

Problems of Nuclear Power Production

B. T. Feld

The problems raised by the rapidly expanding utilization of nuclear power fall into two categories. The first one concerns the adequacy of currently available safeguards, and the improvements which can be readily contemplated, in coping with the problems of preventing the diversion of fissile materials into nuclear weapons production. It seems to me that this is a technically manageable problem, given an international atmosphere in which the nations involved (the so-called near-nuclear powers) continue to regard it in their national self-interest to inhibit the spread of nuclear weapons to states not yet possessing them.

The second category contains problems of a more long-range nature. These are concerned with the situation that is likely to prevail fifteen or twenty years from now. Assuming the current trends toward increasing reliance on fission power continue, a large number of sovereign states will then have in being very large programs of power production utilizing nuclear (fission) reactors, with the concomitantly immense rate of production of fissile materials capable of being utilized for nuclear weapons.

The following considerations may help to provide a reasonable perspective: at the present time, only a few percent of all power production is by means of nuclear fission, and that mainly in the most developed countries. It is now contemplated that by 1985 or 1990 approximately fifty percent of all power

Bernard T. Feld is Professor of Physics at the Massachusetts Institute of Technology.

will be derived from nuclear power plants. It is, indeed, expected that the main source of power for most of the developing nations will come from this source. Now the figures on the rate of power production in various regions of the world (see the paper by F. A. Long, especially Table 1, p. 274) indicate that the doubling time of power production is approximately 10 years in the United States and somewhat less, around 7 years, in the less-developed portions of the earth. For a 7-year doubling rate, the increase in the rate of world power production in twenty years will be by a factor of 8, and considering that the fraction supplied by fission power plants will increase from a few percent to around half (50 percent), the increase in nuclear power production which we may anticipate over the next twenty years is of the order of 100-fold or more.

Such an increase in fission power will present tremendous problems of radiation contamination of the biosphere, of the disposal of the fission-product wastes from fission reactors, etc., not to speak of the tremendously more difficult problem of safeguarding the fantastic quantities of available fissile materials from falling into the hands of internationally antisocial elements (the so-called Mafia problem).

Already now, with even the relatively minor role being played by fission power plants in the power economies of the developed nations, serious problems of radioactive contamination and fission waste disposal are being encountered. Undoubtedly, such problems are solvable at the present level. But there are serious questions relating to the available means for their solution when the levels of fission power production will have been increased by some two orders of magnitude—especially when one considers that much of this increase will occur in relatively less technically sophisticated nations possessing much less of either the experience or the incentives for coping with environmental pollution problems.

(These problems may very well be amenable to technical solution, given sufficient incentives and the requisite attention by experts, which I do not claim to be. For example, if the radioactive wastes could be boiled down to a sufficiently small volume, it might be possible to place them into a rocket which could be shot into the sun. That would provide us with a dumping ground of essentially infinite capacity. On the other hand, there might well be equally effective but more economically and technically attractive schemes for achieving the same end.)

Now, there are essentially three reasons why we are likely to take the fission route for power expansion over the next few decades. The first derives from a combination of historical accident and technological inertia. Immediately after World War II, when fission power was a new and promising light on the horizon, one of the major incentives for its development stemmed from a widespread belief that the reserves of fossil fuels—oil and coal—were either extremely limited or rapidly approaching a stage where their extraction would become prohibitively expensive; and that such supplies of fossil fuels as remained should be husbanded for use in providing the necessary chemical products derivable from them. The rapid development of fission power was generally seen as the necessary means of preserving these valuable (and, believed at the time, limited) resources. In the intervening time, however, two facts have emerged: first, vastly larger, almost unlimited, sources of oil and of natural gas have been and continue to be discovered in many regions of the earth; and second, it has taken much longer than was originally anticipated to develop the techniques of fission power production to the stage where they are economically competitive with fossil power, particularly when the requirement is added that those relatively inexpensive sources of uranium and thorium that are available should be fully utilized in an efficient "breeding" cycle. Hence the original incentive for fission power development has largely evaporated, although it may still be argued that fossil fuel sources, being irreplaceable, should not be unnecessarily squandered in the continuing power expansion. Nevertheless, a great deal of investment in technical resources and manpower, and much first-class technology, has been committed to the development of economically competitive fission power, and this provides a technical momentum which would be extremely difficult to turn off. (The analogy of the drives toward developing a supersonic transport aircraft because we have learned to fly faster than sound may not be entirely inept.)

The second reason which is driving us toward the increased utilization of fission power is the growing realization that the burning of organic fuels is producing serious, irreversible environmental hazards, mainly through the "greenhouse effect"—changes in the absorption in the upper atmosphere of the sun's radiations through the increased CO_2 concentration—which may produce climatic changes. Fission power does not present this hazard; but there is some question, at least in my mind, as to whether the substitution of the radioactive for the greenhouse hazard does not take us out of the frying pan into the fire.

The third and perhaps most compelling reason for going the fission route is that, assuming we accept the arguments against continued dependence on fossil fuels, we have no alternative. It would appear that we are caught between the devil and the deep blue sea.

But there may still be a way out. That way may lie in thermonuclear or fusion power. Fusion power has a number of very distinct advantages over both fission and fossil power. First, the source material, primarily deuterium or heavy water, is readily available in essentially unlimited quantity—namely, in the oceans. Second, the environmental hazards presented by both fission and organic fuel burning are not present with fusion, or at least to a very much lesser degree. True, neutrons are released in the thermonuclear reaction, but provided one does not utilize these for the nucleosynthesis of fissile elements (by absorption in thorium or uranium), these neutrons can be absorbed in specially constructed shields which, although thereby rendered radioactive, would not need to be processed or moved for centuries. The other radioactive by-product is tritium (the heaviest isotope of hydrogen), but this may be utilized in the fusion cycle (see the remarks of Carson Mark, pp. 133–138).

There is, of course, an environmental hazard associated with fusion reactors, and this is the thermal pollution—the release of heat into the material used to cool the reactor and, eventually, into the atmosphere, or into streams and into the oceans. But this pollution problem is common to *all* power sources and may, indeed, be minimized in thermonuclear sources owing to the higher temperatures and consequent greater attainable efficiencies of conversion of heat energy into electricity; furthermore, there exist distinct possibilities for the direct conversion of fusion energy into electricity, which would thereby very greatly ameliorate this problem.

With all these advantages, why are we not now embarked on the thermonuclear route? The answer is that there is one very serious fly in the ointment—we don't yet know how to produce a controlled thermonuclear reaction; indeed, it is still arguable whether a satisfactory solution will be possible on any reasonable time scale, given the inherent difficulties associated with the containment of high-temperature plasmas as well as the necessity for refractory materials, not yet available, capable of functioning at the extremely high temperatures involved. (Of course, there is no question about the possibility of thermonuclear energy—*vide* the hydrogen bomb—but only as to whether it can be produced under controlled circumstances.) Still, recent developments in this field permit a certain degree of optimism, and experts have variously estimated the time required for the successful achievement of controlled thermonuclear power as somewhere in the 15–30 year range (see articles by E. L. Creutz and L. A. Artsimovich in the July 1970 issue of the *Bulletin of the Atomic Scientists*).

So we are left with the following dilemma: Unless we are willing and able to curb the exponential growth of power production in the next 15–30 years, or perhaps even longer, we have no choice but to depend on fossil or fission fuels, or on some combination thereof, with all the attendant disadvantages. The answer—if, indeed, there is any answer—could lie in some kind of compromise. Perhaps the affluent parts of the world would be willing to accept their current levels of affluence as a reasonable upper limit, for the next few decades in any case, and to curb the exponential power growth as far as they are concerned. (Of course, provision could be made for that small expansion of power which would permit them to spread this affluence uniformly over their populations.) During this period, while putting increased emphasis on the research needed to prove out the fusion power possibilities, the developing countries could be encouraged to confine their power growth to conventional (i.e., fossil) fuel cycles. Eventually, when fusion power has become feasible, the power exponential could be allowed to take off again. Such restraint may be too much to hope for in this anarchistic world of ours, but certainly, before we embark irreversibly on the fission route, it behooves us to ask in all seriousness: Is this trip necessary?

Response to Dr. Feld's Remarks: Comments by W. Häfele

Both of Dr. Feld's arguments are more concerned with the general problems of energy production rather than specifically with fission plants. I have gone into the waste problem on the German scale. After a very careful calculation, I concluded that if the fission products are solidified significantly in a reprocessing plant, then the volume required to store the waste for the next 30 years is only 1,000 cubic meters. This calculation assumes that gradually up to 50% or 60% of our energy will be supplied by fission reactors. The waste must be stored under special conditions for five or six years to enable the heat to dissipate. It may then be included in glass or other chemical-compound forms which are both stable and sufficiently heat conductive. The most promising long-term method of disposal is in salt mines. One of the major concerns in storing the radioactive waste is that it does not contaminate water, either ground water or the ocean. The very existence of salt deposits, however, guarantees that there is no movement of ground water in that area. Moreover, the salt mines are large enough and have very good heat conductivity characteristics. They are, therefore, quite sufficient to handle the problem. It is true that the whole process must be done with care, but it is certainly feasible. Therefore I cannot accept the statement that waste disposal would be a significant problem. The cost of this disposal program was calculated at one or two percent of the total cost of electric power. Even if it were increased by a factor of two or four, it would not matter very much.

The second problem was the release of radioactivity and of heat from a reactor. Except for the problem of heat, completely clean plants can be built. Moreover, the dumping of heat is not exclusively one of nuclear power plants, but is a general characteristic of all civilization. It may in fact impose an upper limit on the ultimate growth of civilization. In August of this year an IAEA conference in New York will, for the first time, attempt to compare the potential dangers of fossil plants with the risks involved in nuclear plants and try to make a quantitative judgment.

Although not for the United States, for many countries the development of nuclear power has been the means by which they have learned to manage an overall large development and a technological system. This operational management aspect, which brings governments, scientific communities, industry, utilities, and the public into cooperation for the first time, has a subtle and complex value. For many countries the civilian use of nuclear power has been the main area in which to learn and to establish these systems-management techniques.

Comments and Summary of Discussion on Safeguarding Nuclear Installations

Concerning Safeguards: Comments by J. Guéron

The one and only nonvoluntary system is the Euratom system. Technically it has been accepted by the USAEC, which has agreed to Euratom instead of AEC safeguards for fissile material supplied from the United States to Euratom countries. This fact does not exclude, of course, attempts at improvement. In scope the Euratom safeguards embrace *all* nuclear facilities in the six Common Market countries, except those officially declared as military ones (as per the Euratom Treaty).

Politically it has been argued that such safeguards amount only to self-control. It would be so if the six countries had formed an actual federation. Such is not yet the case. Even in atomic energy the national nuclear programs by far outweigh that of the community (the French as well as the German civilian nuclear R&D programs each are bigger than Euratom's, not to speak of nuclear power plant investment).

Safeguards negotiations between IAEA and Euratom are to be undertaken so as to harmonize the duties of Euratom and nonproliferation treaty obligations (i.e., all six, except France).

Safeguards and Physical Security: Comments by J. Prawitz

I would like to make one point about peaceful nuclear activities. It should be appreciated that the safeguards system now being phased in, in accordance with the Treaty on Non-proliferation of Nuclear Weapons, provides only for information gathering. It cannot prevent the diversion to military uses of safeguarded material, if the host country wants to do so, i.e., it does not provide, and is not intended to provide, any physical security preventing deliberate military use of safeguarded material. It will only detect such use.

It is important for the discussion of the possible impact of the expected buildup of large plutonium stockpiles on the general strategic balance that the functions of information gathering and maintenance of physical security are not confused. The maintenance of physical security, in addition to safeguards, is provided for in many bilateral agreements for international atomic power cooperation and fuel supply. There is, e.g., the supplier's option to buy back plutonium produced from supplied nuclear fuel, if he thinks that the plutonium will be safer in his custody than in the customer's hands. (There are also economic interests built into this provision.) The discussion that has taken place in Europe, and which I think will continue, with respect to the location of a centrifuge facility for U^{235} production partially reflects a concern about physical security. There is no question that this facility will be safeguarded. The problem is to estimate the risk that the facility, sometime in the future, will be turned to military production by the host country. Some people think that location in a nuclear weapon state would best serve the purpose of nonproliferation, while others claim that the risk of diversion to atomic bomb production would be largest precisely there.

I don't want to express myself just now either against or in favor of more physical security, but I felt clarification was justified.

Tritium Safeguards: Comments by P. L. Ølgaard

Professor Long mentions in his paper that it is easier to obtain agreement on disarmament measures in areas where the technological development is still at an early stage than in areas where deployment of weapon systems has already taken place. I agree fully with this view and would, in this connection, like to bring up the question of safeguarding tritium production.

We have heard earlier in this meeting that it may some day be possible to produce thermonuclear weapons which will not be triggered by a fission device but which will instead be ignited by use, for example, of a laser system. Such a development will probably not be of major importance to the nuclear powers since they already possess efficient and rather cheap thermonuclear weapons; but it may have serious consequence for the proliferation of nuclear weapons.

If such weapons turn out to be technically feasible, it is most likely that any industrially developed country would be able to manufacture them. Such countries—and only such countries—should be in a position to perform the required development of compact, high-power laser systems and to undertake the production of relevant thermonuclear materials.

There are many indications that it will by no means be easy to achieve the necessary temperatures in thermonuclear materials by use of a laser trigger. Hence such a weapon would most likely be based on the D-T reaction which has the lowest ignition temperature of the available thermonuclear reactions.

It may be possible to build fission-triggered H-weapons without an initial tritium content, even if the weapon is mainly based on the D-T reaction. The reason is that the fissions may, in addition to heating up the thermonuclear materials, be able to produce sufficient amounts of tritium, by capture of fission neutrons in Li^6, to start the thermonuclear chain reaction. However in a laser-triggered H-weapon no such neutrons exist and consequently such a weapon must contain deuterium, and probably lithium, enriched in Li^6.

It is important to note that all these materials, tritium, deuterium, and enriched lithium, can be produced without violating the nonproliferation-treaty safeguards, since these deal with fissile and fertile materials only.

Production of such weapons by a nonnuclear country which has signed the nonproliferation treaty would of course mean a violation of the treaty. But it is likely to be very difficult to prove that a country has embarked on production of such a weapon. Research projects exist today which are directed towards the use of laser-triggered thermonuclear reactions for peaceful power production and the techniques used here are very similar to those used in weapon development. Hence any country working on high-power lasers and production of thermonuclear materials can claim that its efforts are directed only toward peaceful uses. It should also be possible to test such weapons on a very small scale, and this would make detection of tests very difficult.

One possibility is to control or prohibit the development of high-power lasers in the nonnuclear countries adhering to the nonproliferation treaty. Such an approach is hardly politically possible. The nonproliferation treaty is already rather unbalanced in favor of the nuclear powers. Any attempt to make it even more unbalanced, by introduction of controls or prohibitions in areas of the technology which may later turn out also to be important in fields with no relation to weapon production, are not likely to be acceptable to the nonnuclear countries if applied, as is probable, only to them. It may even jeopardize the nonproliferation treaty.

Another possibility is to safeguard the production of thermonuclear materials. It is likely to be difficult to safeguard the production of deuterium and lithium, including enriched lithium. A number of countries have plants for the production of heavy water and a much larger number of countries possess significant amounts of heavy water in the form of reactor moderators. Lithium deposits exist in many places and lithium enrichment plants are small and relatively easy to hide.

However, as discussed above, tritium seems to be essential to the production of non-fission-triggered H-weapons and safeguarding tritium production should be relatively simple. The reason is that tritium is produced today in fission reactors which are already safeguarded. Should fusion reactors prove feasible, they will also produce tritium, and hence they will also have to be safeguarded. Since they could also be used to produce fissile materials, they would probably have to be safeguarded anyway.

Consequently, it seems very desirable and also technically feasible to extend the IAEA safeguard to include tritium so as to obtain adequate ensurance that non-fission-triggered thermonuclear weapons are not produced by nonnuclear countries adhering to the nonproliferation treaty. It should be fairly easy and simple to accomplish this extension of the present safeguards system, although it may demand a change in the statutes of the IAEA.

It may be said that the question of tritium safeguard is not very urgent today. However, if one waits too long to introduce tritium safeguards, it may be too late.

Summary of Discussion

International Atomic Energy Agency Safeguards. The view was expressed that the present IAEA safeguard provisions are sufficient for immediate implementation. Dr. Häfele was asked to comment on the compatibility of the safeguard procedures he described and those of the International Atomic Energy

Agency, IAEA. His comments included the following statement:

"It is my feeling that the IAEA document is a useful framework to stand on but is by no means a recipe for pursuing any kind of safeguard. The document is very broad and many kinds of safeguard procedures could be implemented and still stay within its limits. What is under debate is a more detailed description of safeguard procedures. The real point, therefore, is not whether the final safeguard procedure is within the document, but rather the intrusiveness or the nonintrusiveness of the procedure.

"There is one particular aspect of the IAEA document which makes it unfeasible for immediate implementation. When the document was conceived, it was the single, isolated nuclear facility that was to be safeguarded. But actually the whole fuel cycle must be safeguarded, making use of its interdependences."

Material Unaccounted for. It was pointed out that with effort and planning, government officials in Great Britain were able to steal one kilogram of plutonium. This raised the more general question of MUF, material unaccounted for, which can be of the order of one to two percent of the total fissile-material inventory.

Dr. Häfele's remarks on the MUF question were as follows:

"Despite twenty-five years of general operational experience in a number of reprocessing plants, the question of MUF has not been fully analyzed, and it is therefore not possible to say how large it really is or what its source is. Recently, a number of well-targeted experiments have been carried out, of which our group has done several. Originally we felt that the statistical error would be important, but it turns out that the systematic error in making the principal input and output measurements dominates. Despite previous doubts, we managed to reduce MUF to a level of two grams in the plutonium stage of the reprocessing plant that we studied. The point is that MUF is only now being understood and it is therefore not possible to extrapolate from the past.

"It is true that MUF will always exist at some level, but it need not build up over time, but rather could remain constant. It is not possible over a period of years to lose a ton of plutonium."

Proliferation of Nuclear Technology. From the point of view of the major industrial countries there is a trade-off between the dangers of small diversions of fissionable plutonium from nuclear power plants and the economic advantages of the rapid spread of nuclear reactor technology to small countries. In their pressing need for increased power sources, small countries may not see the problem in the same light.

It was pointed out that the industrial users of nuclear power cannot be expected to await the final decision on safeguard procedures. Rather the safeguard proposals must keep pace with the growing industrial use.

Non-bomb Military Use of Fissile Material. The non-bomb military use of fissionable material, particularly for submarine power sources, is not covered by the nonproliferation treaty. There is some movement to plug this gap in the safeguard requirements and thereby to make the final safeguard agreements more stringent than is required by the nonproliferation treaty. This area has not yet been given sufficient attention.

Supercentrifuge Technology. The nonproliferation treaty allows the use of any and all fissionable material in the civilian domain, including the possibility of highly enriched uranium. There is, therefore, no reason why a centrifuge plant should not be used for industrial purposes. In practice, however, the enrichment would probably go only to three or four percent because, except for very special and rare applications, this is all that is necessary. Rather than trying to prevent the use of this new technology, sufficient and proper safeguards should be developed for such plants.

There are certain aspects of centrifuge technology which give it a marginal advantage over diffusion separation if one were interested in building and operating a clandestine plant. The centrifuge does not require the very high input of power that a diffusion plant demands. A centrifuge plant is not restricted nearly so strongly by the economies of scale that govern a diffusion plant, and therefore could be built to have a very much smaller throughput. Furthermore, it is reasonably easy to rearrange the configuration of a centrifuge plant, so that a facility originally planned for relatively low enrichment could easily be transformed into one of very high enrichment. On the other hand, to maintain concealment of these facilities over long periods of time would be rather difficult, especially in the face of traditional intelligence gathering which countries are in the habit of practicing.

3 Military Research and Development

Growth Characteristics of Military Research and Development

F. A. Long

Overall Growth of Science and Technology

This analysis is principally concerned with the growth characteristics of worldwide military-oriented research and development (R&D). Its main objective is to develop understandings that will permit more effective civilian control of military technology. Military technology and the R&D that lead to it clearly have some special characteristics. Nevertheless, they must be thought of as only one component of the overall development of science and technology. We therefore start with a discussion of the growth of the total science and technology effort.

Analysis of the scientific effort of the world clearly shows that a rapid, indeed exponential, growth has occurred. Figure 1, taken from de Solla Price's book, *Science Since Babylon* (Ref. 1), is a good illustration of this. Price persuasively argues that the curve of this graph is a true exponential, only displaced twice by the consequences of two great wars. Now, everyone knows that this sort of exponential growth cannot continue indefinitely, and Price himself discusses the implication of the shift over to a sigmoid curve, which would be the result of a trend toward saturation of science. However, at this point in time, we are clearly still on the exponential curve; and even if some of the developed nations soon show signs of some leveling off, this will almost surely not be true for the

Franklin Long was formerly Assistant Director of the United States Arms Control and Disarmament Agency. He is now Professor of Chemistry at Cornell University.

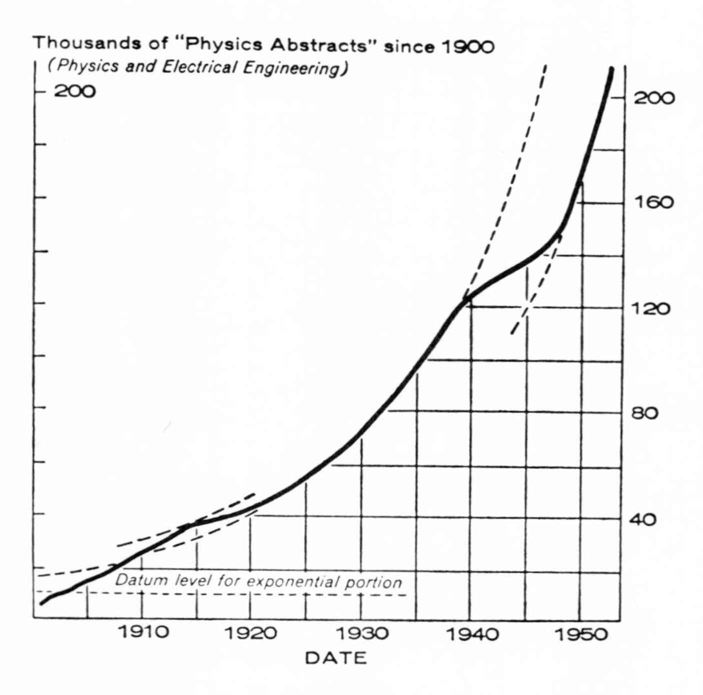

Figure 1
Total number of physics abstracts published since January 1, 1900. The full curve gives the total, and the broken curve represents the exponential approximation. Parallel curves are drawn to enable the effect of the wars to be illustrated.

developing nations. Indeed, Price is at his speculative best on this aspect of the analysis. He has produced an ingenious set of curves (see Figure 2), which essentially argue that the laggards in the development of science can build on the results of the leaders to the point not only of catching up to, but

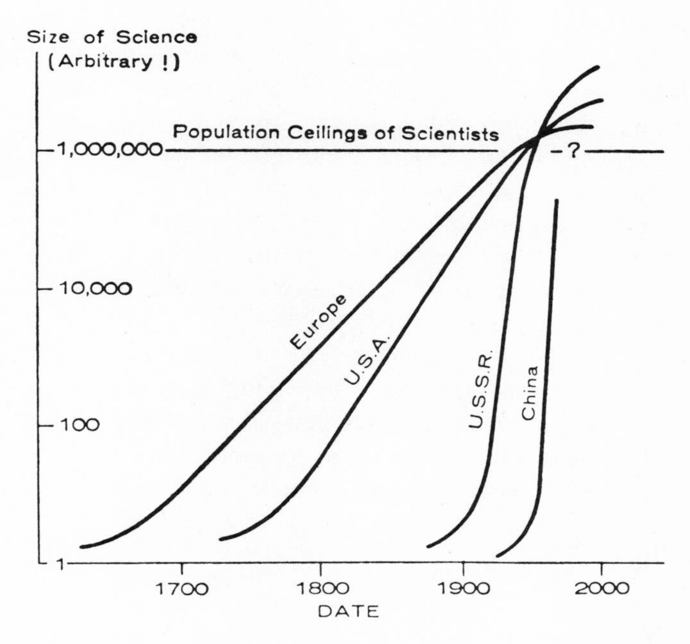

Figure 2
Schematic graph of the rise of science in various world regions. The measures, the shapes of the initial portions of the curves, and the way in which the curves turn over to their respective ceilings toward the top are all merely qualitative.

perhaps going beyond the former leaders. "Expanding into a scientific vacuum" is the catchy phrase Price uses to describe this situation (Ref. 2). Here, if you will, is still another aspect of the China threat—what a great vacuum they have to expand into! While these illustrations relate to the growth of science, not of technology, I think there have been sufficient analyses of the direct flow of ideas from science into technology to make evident the close connection between the two. The example comes to mind of the fine analysis of the link between the science of chemistry and the chemical industry, to be found in the 1965 National Academy publication, *Chemistry: Its Opportunities and Needs* (Ref. 3). (Incidentally, the general public is well persuaded that there is a close link. People speak of "science and technology" as though it were a single subject.)

It is a truism—perhaps one that is too easy—that ours is an increasingly technological civilization; but there are many possible measurements that demonstrate this. Let me give you only one—the growth of electrical consumption in the United States and in the world (Ref. 4). Table 1 gives details.

Table 1 Electric Power Production (kw hours $\times$ 10^{-9}).

Year	U.S.	World
1902	6	—
1920	57	—
1930	115	271
1940	180	482
1950	389	855
1960	842	2,304
1965	1,158	3,844

Similar trends hold for automobiles, telephones, pharmaceuticals, commercial air travel, and, no doubt, for potato chips. Technology has clearly grown to the point where, in many nations, it and its fruits dominate daily life.

We can agree, then, that both science and technology have expanded enormously in recent years. To produce this great growth has called for trained people—specifically, scientists, engineers, technology-oriented managers, and the like. And, indeed, such people have been produced in sharply increasing numbers. Table 2 gives two illustrations of this, one for the United States (Ref. 5), and one for the USSR (Ref. 6).

Table 2 Illustrations of Growth in Personnel.

A. Production by decade of U.S. Ph.D.'s in Natural Sciences:

Decade	Number
1920–29	5,864
1930–39	12,604
1940–49	15,502
1950–59	40,089

B. Personnel employed in science and science services in the USSR:

Year	Number
1940	361,000
1950	714,000
1960	1,765,000

This great growth of science and technology and of scientists and technicians has been paralleled by the growth of a set of industries skilled in the sophisticated application of technology. The central theme of one of the most important books in recent years, Galbraith's *The New Industrial State* (Ref. 7), is that this new and growing segment of industry relies heavily on science and technology, is committed to growth, and in general has characteristics that are in most ways independent of the ideology of the national governments under which it operates. From our standpoint, the most significant characteristic of this industry is its steadily increasing skill in converting basic and applied science into usable technology.

Most scientists and engineers are convinced that the time needed for industry to convert scientific discoveries into practical technology has steadily shortened in recent years. Firm statistical support for this conviction is not easy to obtain, but the example of the great Apollo program makes the point dramatically. As we all know, the intricacies of the Apollo task—in President Kennedy's words, "A manned lunar landing and safe return in this decade"—and the technological sophistication necessary to solve the problems that faced it were almost beyond belief. Even so, at a cost of some 30 billion dollars plus, the Kennedy goal was fully accomplished.

This example also illustrates a very different but also important point: the extensive interaction that often exists between technically oriented governmental agencies and the groups that develop and produce their technology. Although most Western countries make much of the distinction between private enterprise and government, as Galbraith points out: "the line between public and private authority in the industrial system is indistinct and in large measure imaginary" (Ref. 8). With respect to the Apollo program we were just discussing, this is surely an apt characterization of the relations that have existed between the National Aeronautics and Space Administration and the many private companies engaged in development and testing of the Apollo equipment.

To summarize these preliminary points, we can say the following:

1. For many decades, there has been a world-wide, very rapid exponential growth of science and technology. A consequence of this is the technological civilization which characterizes the developed nations of the world.
2. Although a trend toward saturation in scientific effort may soon be encountered by the developed nations, it is less clear that it will soon occur in technology. In any event, for the larger fraction of the world's population the great growth in science and technology is yet to come and, when it does come, it will probably be explosively rapid.

3. Along with the growth in the number of scientists and engineers there has come a sophisticated technology-based industry adept at applying science and technology to the world's needs and desires.

It is within this general context that we now turn to the characteristics of military technology and the R&D that produces it.

Development of Military Technology.
Military technology is concerned, among other things, with trucks, tanks, and ships, with guns, bombs, and explosives, with rockets and airplanes, with surveillance and communication. Even this brief list illustrates the substantial overlap between civilian and military technology. Much of the development effort that leads to new military technology is an outgrowth of, or is done in parallel with, development efforts for new civilian technology. Indeed, in many technical areas—computer development, for example—the efforts directed toward military and civilian technology are quite indistinguishable.

It is also true that military and civilian technology ultimately stem from the same basic and applied science. Transistors and similar solid-state devices have largely replaced vacuum tubes in both military and civilian communications systems. Furthermore, the ultimate production of sophisticated military equipment necessarily involves the same kinds of production-line facilities, specialized equipment, and skilled workers and managers that much civilian technology involves. In a quite fundamental sense, then, military R&D is embedded in and derivative of civilian science and technology. Finally, one must note that there is often a reverse flow or "spin-off" of products and procedures from military R&D into civilian technology. Extreme proponents of large military R&D efforts have occasionally attempted to justify these efforts by pointing to the spin-off benefits alone.

These important areas of overlap and interdependence do not, however, mean that there are not substantial differences between civilian efforts and most of military R&D. Their essential goals are clearly different. Most civilian technology is directed (or should be directed) toward supplying desired civilian goods and services, and improving the quality of life. Military technology is directed toward military defense and toward success in war. A successful nuclear deterrent, hopefully to prevent a major nuclear war, is a key goal; survival and ultimate success on the battlefield of a conventional war is another. These goals determine the character of the products needed to reach them. Much military technology is characterized by the production of specialized products, solely designed for war. The battlefield tank, for example, is a unique and specialized device imaginable only in the context of war. The same may be said of intercontinental ballistic missiles and nerve gas.

A more detailed statement of goals and needs of

military technology comes from a recent analysis by Dr. Donald MacArthur, Deputy Director for Research and Engineering of the Department of Defense (Ref. 9). His list of needed continuing improvements, somewhat condensed, is:

First . . . emphasis on all of the equipment required for a sufficient and credible strategic nuclear deterrent. . . .
Second . . . improve our *all-weather all-climate* fighting capability. . . .
Third, high reliability and greater flexibility. . . .
Fourth, mobile and flexible deployment systems in small units. . . .
Fifth, much better understanding of the relationship among the military, political, economic, technical, and psychological factors influencing successful deterrence . . . of the use, or the threat of use, of force.
Sixth, strategic and tactical intelligence and surveillance data collecting and processing systems.
Seventh . . comprehensive command-control communications systems. . . .

These specialized goals clearly require developmental groups that are distinct from those concerned with civilian technology. Such groups exist in large numbers throughout the world. In the United States, several scores of them are to be found in the Federal Government, mostly as laboratories or arsenals of agencies of the Department of Defense. Other large groups in the United States are located in those industrial corporations that specialize in work for the military.

Some idea of the magnitude of the trained personnel and special facilities in the U.S. effort of military R&D is given by the following facts. In fiscal 1969, of the Department of Defense (DOD) funds used for Research, Development, Test and Evaluation (RDT&E), federal laboratories spent $2.4 billion and the amount spent by industry was more than twice this, $5.1 billion. Second, the overall industrial effort for military technology in the United States is of such magnitude that at least four of the corporations that devote themselves principally to technology programs for DOD and NASA (including, for example, McDonnell-Douglas and Boeing Aircraft) are among the twenty-five largest companies in the United States.

Another characteristic that sets military R&D apart from civilian technology efforts is the secrecy which normally surrounds it, especially at the level of development and application. There are strenuous debates as to whether the full amount of secrecy that characterizes military development is necessary or desirable. But the fact remains that an extensive blanket of secrecy is a worldwide characteristic of military R&D, and this situation is not likely to change. Two immensely significant consequences follow. One is that this "security fence" of secrecy acts as a barrier to informal interchange of ideas between the scientists and engineers involved in military research and their civilian counterparts. A second is that this same security fence seriously inhibits application of the same degree of examination and assessment of the military R&D effort by citizens and legislators that usually characterizes large and expensive government programs. This last point is an important one, which ought perhaps to be emphasized by one or two comparisons. Annual expenditures by the U.S. Department of Defense for military RDT&E have, in recent years, been somewhat in excess of 7 billion dollars. In the same period, the NASA annual budget, principally for the Apollo Program, has varied from $3 to $6 billion. But the NASA Program has been the subject of far more books, articles, newspaper editorials, and congressional hearings than has the budget for military RDT&E. It is true and important that the budget for the military R&D effort is to a considerable degree "shielded" by the total military budget. But even at the total budget level, there is a great difference in the degree of analysis. Compare the debate, analysis, and congressional hearings that are devoted to U.S. federal expenditures for education with the much smaller discussion that is directed toward the considerably larger federal expenditures for the military.

Expenditures for Military Research and Development

It is difficult to obtain good expenditure figures, even for the total military efforts of the world. It is far harder to obtain good analyses of expenditures for military R&D. Because the United States has for some years been publishing fairly complete data on most of the agencies concerned with the military, one tends to utilize the U.S. data as illustrative. It is reasonable to assume that similar percentages of funds for the military go into R&D for all of the developed nations of the world. But firm evidence is lacking.

According to the U.S. Arms Control and Disarmament Agency (Ref. 10), world military expenditure in 1969 reached the staggering sum of $200 billion, up 44% from 1964. The two major powers, the United States and USSR, spent roughly $80 and $55 billion, respectively, about two-thirds of the total. For the United States, and presumably for the other developed nations, funds for RDT&E were about 10% of its total military budget. Specifically, for the United States the recent budgets of the DOD have included about $7 billion per year for military RDT&E.

In considering the magnitude of expenditure for military research and development, it is important to realize that all the funds for this effort are by no means to be found under the rubrics "military" or "defense." For example, in the United States there are three significant categories of military-related R&D which are managed by agencies other than the Department of Defense. One is R&D performed by the Atomic Energy Commission on development and testing of nuclear weapons. In 1964, the AEC spent $1.2 billion for R&D, and it is reasonable to estimate that more than half this sum related to nuclear weapons. Much of the R&D done by the Central

Intelligence Agency almost surely is closely related to military intelligence efforts but is budgeted apart from the DOD. Finally, many of the R&D programs of NASA in such fields as, for example, communication, surveillance, rocket guidance, and data processing are closely related to areas of military interest. Indeed, the statute which established NASA requires that explicit attention be given to defense applications of space programs. One cannot be quantitative about the amount of military R&D done by these other U.S. agencies, but an estimate of an additional $2 billion per year seems conservative. Presumably in other nations a similar interrelationship among agencies exists for the performance of military R&D.*

Two aspects of the growth of DOD-sponsored military RDT&E in the United States are shown in

* Evidence on both these last points is to be found in a recent talk by Dr. John S. Foster, Jr., Director of Defense Research and Engineering (given at the 16th Annual James Forrestal Award Dinner, National Security Industrial Association, Washington, D.C., March 12, 1970). In discussing comparative levels of effort in the United States and USSR, Dr. Foster said: "Let us now move from the overall national picture to the defense-related R&D. Here we must include atomic energy and space as well as the work most narrowly focused on the armed forces. This year, the Soviet Union is investing the equivalent of 16 to 17 billion dollars in such defense-related research, development and applications. The United States is investing 13 to 14 billion dollars in comparable activities." It seems clear from this that for the United States, the contribution of non-DOD efforts to the total defense-related research effort is almost as large as that of DOD itself. Also of interest is this DOD estimate that the military R&D effort of the USSR is as large or larger than that of the United States.

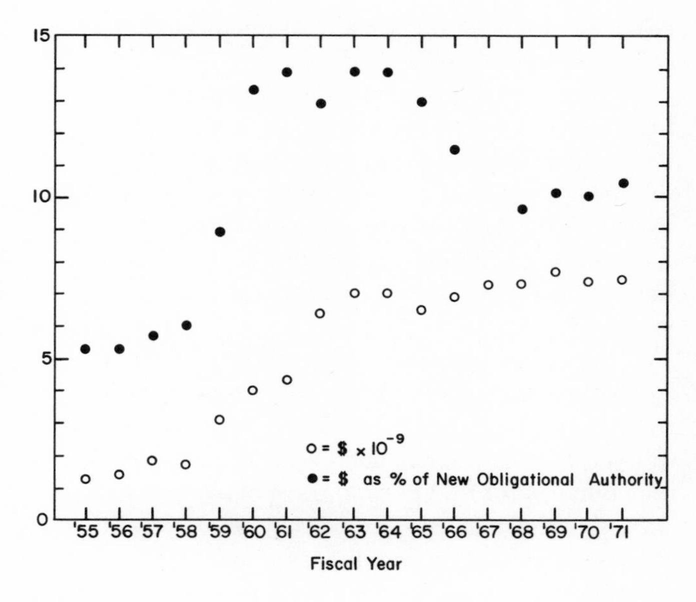

Figure 3
U.S. expenditures of DOD funds for research, development, and test and evaluation. Figures for 1971 are as proposed.

Figure 3. One is the absolute growth of these funds; the second is their percentage of "new DOD Obligational Authority," i.e., the funds available for new things. Up to recent times the funds for RDT&E have grown steadily and substantially both in absolute dollars and in percentage of new authority. It is notoriously dangerous to extrapolate curves like those of Figure 3, but at the very least it seems safe to say that in the absence of a sharply different climate in foreign affairs or of new procedures for civilian assessment and control of the military, these expenditures will grow.

In considering this point, an impressive fact from Figure 3 is that even under the burden of turning over large funds to the military efforts in Southeast Asia—amounts which have ranged upward toward $30 billion per year—the DOD has been generous with RDT&E. Funds for RDT&E have not declined and have, in fact, grown slightly in recent years.

The more detailed analysis of Figure 4 helps explain this fact and helps support the prediction that accelerated growth for RDT&E expenditures is in the offing. This figure subdivides the total DOD expenditures for RDT&E into three categories which vary from the relatively longer-range research and development to short-range operational-systems development. Funds for the longer-range efforts (Curve A) have been virtually constant for several years. There was an upswing in these funds in the early 1960s, followed by a slight drop in the last few years, perhaps as a result of pressures for operational funds for the Vietnam war. Funds for operational-systems development (Curve C) increased steadily through 1969 but have dropped sharply in the last two years. This too may relate to Vietnam pressures, although in part it may simply reflect the final full deployment of some of the U.S. strategic systems which were started in the early 1960s. Advanced and engineering development (Curve B), after holding steady for several years at just under $2 billion,

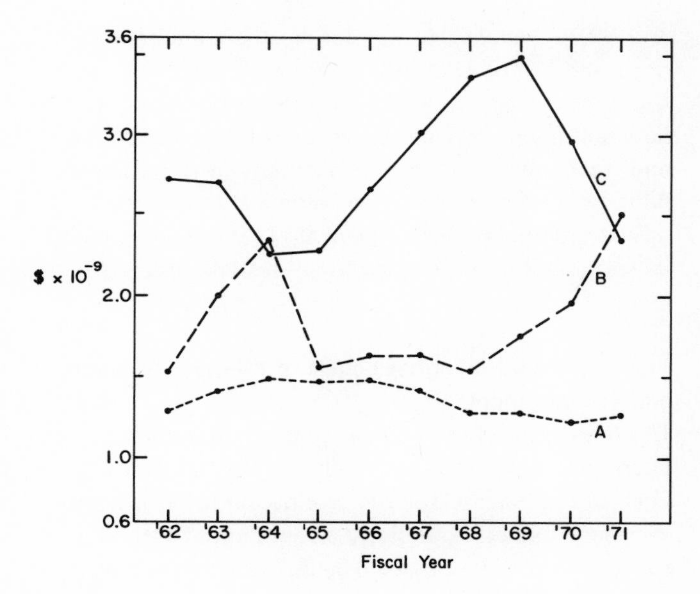

Figure 4
DOD expenditures (in current dollars) by fiscal year for three segments of RDT&E: *A*, research plus exploratory development; *B*, advanced development plus engineering development; *C*, operational-systems development. Figures for 1971 are as proposed.

appears to be exhibiting a sharp increase currently. An implication of this increase is that some significant new systems are being developed, and that a buildup of funds for operational-systems development is likely to follow in a year or two.

As further indication of what is ahead, we do know that two major new U.S. developmental programs are now very much underway, the ABM and MIRV systems. A number of others are being proposed by the U.S. military for early decision. Among them are an advanced manned bomber, a new submarine missile carrier, new tactical aircraft, and new helicopter systems. Agreement to go ahead with even a few of these will almost surely push upward requirements for new funds for engineering development and for operational-systems development.

Pressures toward Higher Levels of Military Research and Development

There are a number of forces that simultaneously push military R&D to ever higher levels of effort. First, of course, there is the worldwide trend toward increased use of technology as an alternative and supplement to manpower. The military does not escape this trend. Dr. MacArthur of the Pentagon concluded, in the article referred to earlier (Ref. 9), by saying:

In the next 10–20 years, there will be no decrease—in fact there will probably be an increase—in the strong dependence of national security upon advanced [military] technology.

A second source of pressure is the rivalry among nations, which almost inescapably tends to convert military R&D efforts into a technology-based arms race. If the USSR develops an improved missile system, the U.S. military feels impelled to respond. When Israel obtains Phantom airplanes, the United Arab Republic presses to obtain a technically viable counterweapon. This spiraling aspect of the arms race was vividly spelled out by Herbert York, former Director of the Office of Defense Research and Engineering of the DOD, in recent testimony to a Senate Committee (Ref. 11). After describing the U.S. and USSR decisions on the ABM and MIRV systems, he summarized:

We thus see that the whole process has made one full turn around the [arms race] spiral: Soviet ABM led to U.S. MIRV; U.S. MIRV led to Soviet MIRV; Soviet MIRV leads to U.S. ABM.

A third more specific source of pressure is a consequence of the momentum of a large program of military technology. The United States, the USSR, and most other developed nations now have an "establishment" of committed laboratories, trained engineers, and large industries, all specializing in the production of military technology. These can exert a very real set of pressures and it appears that President Eisenhower, in his famous 1961 warning against the "military-industrial complex," had in mind the pressure from this kind of establishment.

The particular pressures exerted are both technical and political. They are scientific and technical because these large technology-oriented "defense establishments" cannot fail to generate persuasive proposals for new weapon systems and for improvement on older ones; they are economic and political because the large production efforts of the major defense-oriented industries have become important and accepted means of livelihood in many communities. As an example of the latter point, Boeing Aircraft is the largest industrial employer in the State of Washington, and Boeing's success or failure at maintaining full employment is a matter of deep concern to the senators and representatives from Washington.

An excellent set of illustrations of technical pressures toward more military R&D is to be found in Chapter 2 of the 1968–69 *SIPRI Yearbook of World Armament and Disarmament* (Ref. 12). This discussion gives four examples of the application of R&D to military systems. Each example describes specific ways in which military R&D has been applied. Thus, the analysis of the development and later improvement of the U.S. Polaris system for submarine launched missiles shows in quite a specific way the reasons for and expected benefits from the intensive R&D program which has continued for this weapon system. Incidentally, this SIPRI analysis is also a fine illustration of how much detailed information about a "classified" military system can be obtained from the open literature by a persistent investigator.

A different and disturbing general exposition of reasons why expenditures for military technology tend to remain high in all industrial states is given by J. K. Galbraith (Ref. 13), who argues that for a modern industrial economy to remain on an even keel, there must be continued action by the state to stabilize what the economists call "aggregate economic demand." Galbraith states that to accomplish this there must be large, controllable expenditures of the central government. In addition, there is a real need for state assistance in underwriting new technology. Military expenditures meet both these needs. They have, in the past, also had the added advantage of quasi-automatic public acceptance. The question is, must this continue? Galbraith goes on to propose that for reasons of genuine national security, the United States should diminish its reliance on military expenditures for these purposes and locate other technically sophisticated areas worthy of large-scale federal support, for example, space exploration and pollution control. The assumption is that these, or others like them, would be accepted as appropriate alternatives for commitment of the large federal expenditures needed to manage aggregate demand. One hopes, with Galbraith, that this is true.

Let us summarize. International competition, local and internal political pressures, the postulated needs of the military, and the general thrust of developing science and technology all work to push military R&D to higher levels. What, if any, are the countervailing forces?

Control of Military Research and Development

As a preliminary comment, it must be emphasized that control of military R&D *by itself* appears to be peculiarly difficult. In the first place, the question of what is "military" research and development is often ambiguous. Second, military R&D is generally treated as more innocuous than actual production of military goods. Finally, one notes there is a tendency to argue that, in the context of arms limitations or disarmament, military R&D should in fact be *increased*, presumably in order to minimize the possibility of some "technological surprise" (Ref. 14). Thus, although some specific aspects of military R&D may be subject to direct and effective control, it is more likely that control of total military R&D will be effected through control of the total military effort.

A discussion of mechanisms for controlling military R&D can benefit from a categorization of possible strategies of control. It will be useful to consider:

1. A piecemeal approach to selected, individual R&D programs, i.e., "chewing off bits and pieces of the problem."
2. Control of broad segments of RDT&E, which in a sense can be thought of as "control by redefining the problem."
3. An attack on the total control problem. This might conceivably be directed toward military R&D itself. But it is more likely to be effective if directed at overall military expenditures. One can usefully subcategorize this type of approach into "budget control" and "citizen control."
4. Broad international agreements developing out of the perspective of item 3 above.

Piecemeal control of military R&D will usually be associated with the restraining effect of specific measures to control or, much more preferably, to prohibit, certain types of military weapon systems. Thus the signatories of the nonproliferation treaty will have little incentive to continue a major R&D effort on the development of nuclear-weapon systems. Similarly, the international treaty on "No Bombs in Orbit" has presumably substantially diminished R&D programs for the development of orbiting military weapons-delivery systems. In a different area, the recent proclamation of President Nixon banning any U.S. production or use of biological weapons will almost surely decrease the R&D effort of the United States in this particular area. All of these are encouraging steps, but as a control mechanism one must view this piecemeal approach with somewhat mixed feelings. Each act of control or restraint can surely be regarded as a plus, particularly if it has an international dimension. At the same time, one must be sharply aware that new science and technology is steadily bringing along new weapon-system possibilities. Hence, one cannot expect a piecemeal approach to be very successful in achieving any total control of military R&D.

The subcategories utilized in the analysis of R&D expenditures of Figure 4 suggest a procedure whereby some control of RDT&E may be attained at the national level—and perhaps also the international—by giving appropriate attention to the various categories of effort. The essential point of Figure 4 is that there is a spectrum of R&D activities, varying from quite long-range research and exploratory development to short-range operational-systems development. When it is argued that national security requires continued military R&D, even in context of agreed arms-limitation treaties, the usual implicit meaning is that what is needed is a continued program of long-range research and exploratory development. This is the kind of program which can presumably minimize the possibility of technological surprise. On the other hand, it is the more immediate development programs, e.g., on weapon systems and on advanced engineering, which tend to be more expensive and which also generate the greatest momentum toward actual systems deployment. It is therefore conceivable that careful discrimination among the different categories of R&D, focusing control efforts on the activities closest to deployment, may permit more efficient and useful control of those activities, even in the context of maintaining a vigorous long-range research program. Incidentally, the advanced stages of development also represent the stages that are generally most "visible," and hence, in context of an international agreement, more likely to be accessible to unilateral verification. The essential conclusions one can draw, then, are that there may be many more promising and realistic possibilities for control of R&D at the systems-development end of the R&D spectrum than at the research end; and furthermore, that this may be the more desirable approach. However, as with the previous category of control measures, this does not appear to be an adequate procedure for attaining overall control.

As I have suggested, it seems probable that control of overall military R&D at either the national or the international level will be most effectively obtained by control over total military expenditure—in other words, budget control. No nation has the resources to pursue at requested levels all the programs that groups of its citizens plead for. Choices must be made and priorities set. The complex negotiations that lead to an acceptable national program are, in most countries, summarized in the agreed national budget. The military must argue its case in this budget-making forum.

Regrettably, the persuasiveness of the military in these priority discussions is traditionally very high. It has been all too easy for them to persuade large numbers of citizens and legislators that "national security" should be equated with military strength. Furthermore, the existence among nations of harsh international competition—essentially based on this idea that national security depends on military strength—has been strongly supportive of large military budgets. Thus, it is customary for the U.S. military, seeking support for their budget requests, to paint in grim words the advanced state and rapid development of Soviet technology. No doubt a similar approach is used by the USSR military. The problem then is to find ways to exert meaningful budget control in the face of this traditional persuasiveness of the military.

It is particularly important that the legislative branches of government, in their representative role, carry out a continuing critical analysis of military programs and budgets. To use the United States as an example, virtually the entire consideration of military affairs by Congress is limited to the two armed-services committees, which, over the years, have earned a well-deserved reputation as committees that are friendly and sympathetic to the military and that are easily persuaded of the utility of new military programs and of the centrality of the military in all aspects of national security affairs. We must change the situation in which, in the words of Senator Gaylord Nelson, it is

An established tradition . . . that a bill to spend billions of dollars for the machinery of war must be rushed through the House and Senate in a matter of hours, while a treaty to advance the cause of peace or a program to help the underdeveloped nations or . . . to advance the interests of the poor must be scrutinized and debated and amended and thrashed over for weeks and perhaps months. (Ref. 15)

One of the unhappy consequences of the rise of military activity in recent years has been the worldwide movement of military agencies into positions of decisive influence in areas that go well beyond the narrow sphere of military problems—for example, into foreign affairs and programs of foreign economic development. Ways need to be found to restore to the state departments and the foreign offices of the world a position of centrality and control in these areas and to give them, as well, a strong voice in national security decision that, though they may appear so to the military, are not entirely of a military character.

As a possible solution to this problem, as well as a contribution to firm budget control, a joint legislative committee for national security affairs might be established, formally charged to hold hearings and conduct critical analyses of national security in the broadest sense. The responsibilities of such a committee should by no means be limited to the military but should include consideration of programs of economic aid, regional cooperation, international peacekeeping, and arms control and disarmament as alternative pathways to national security. Here is a program of control that the United States and other nations can develop unilaterally and immediately.

More generally, there must be much more citizen knowledge of and participation in solving problems of national security viewed broadly. This vital area must not be the exclusive province of the professional scholars and the military. The universities have been negligent in teaching students and training scholars in this area; they must be persuaded to change their ways. Foundations and other philanthropic groups must lend support to more, and more substantive, discussions and analyses. The recent public debates and discussions within the United States on the ABM issue, and within West Germany on relations with Eastern Europe, are encouraging signs. There need to be many more such debates. But beyond these nationally based efforts, and perhaps more essential for the attainment of world peace, are international agreements for control of the military and for avoidance of war. To these we now turn.

International attempts to develop controls of the level of military activity have often focused on proposed formal agreements for parallel reductions in national military budgets. Specific measures to effect this have frequently been proposed at international conferences but have usually been rapidly dismissed by one nation or another as unfeasible. Nevertheless, this type of agreement probably merits serious reexamination. With world military expenditures now at the level of $200 billion a year, the pressures for recapturing much of this for important civilian programs must be almost universal. A principal argument against formal international agreements on restrictions of military expenditures has been the conviction that such an agreement could be too easily avoided by simply defining certain sorts of military effort as "nonmilitary." Granted that this sort of avoidance is clearly possible, there are still reasons why this kind of agreement may nevertheless be useful. One is that modern surveillance methods make it less likely that nations bent on violation could successfully conceal major military buildups, even if they were carried out under some kind of "cover" program. Another reason is that the desire and need to be relieved of the heavy economic burden of the arms race will constitute a strong incentive for continued acceptance of this kind of limitation.

It remains true, however, that the best hope for international control of military efforts is most likely to be found in broad agreements to limit armaments and to provide better peacekeeping machinery for the world, i.e., to provide mechanisms by which nations, in their search for true national security, can confidently decrease their reliance on the military. Success in SALT, for example, could greatly decrease U.S. and USSR pressures toward new strategic-weapon systems and new strategic defense systems. If a substantive agreement on strategic nuclear weapons could be negotiated and could be followed by the development of more effective United Nations procedures for peacekeeping, the impact on total military expenditures as well as on military R&D could be major.

One thing is certain. To contain and decrease the spread of military technology and the growth of military spending calls for positive action at both the national and international level. If we stand passively by, the many strong pressures toward more and more expensive military technology will surely prevail.

References

1. D. J. deSolla Price, *Science Since Babylon,* Yale University Press, New Haven, Connecticut, 1961, p. 103.

2. *Ibid.,* p. 111.

3. *Chemistry: Its Opportunities and Needs,* National Academy of Sciences, Washington, D.C., 1956, chapter 2.

4. *Statistical Aspects of the U.S.,* Bureau of the Census, Department of Commerce, Washington, D.C., 1969; W. S. Woytinsky, *World Population and Production,* Twentieth Century Fund, New York, 1953.

5. Data from Office of Scientific Personnel, National Academy of Sciences.

6. *The Research and Development Effort,* U.S. Organization for Economic Cooperation and Development, Washington, D.C., 1965.

7. J. K. Galbraith, *The New Industrial State,* Houghton Mifflin, Boston, 1967.

8. Signet paperback edition of Galbraith, p. 304 et seq.

9. D. MacArthur, *Defense Industry Bulletin,* October 1969, pp. 21–24.

10. *World Military Expenditures,* U.S. Arms Control and Disarmament Agency, Washington, D.C., 1969. See also analysis by R. F. Kaufman, *New York Times Magazine,* June 22, 1969, p. 10.

11. H. F. York, Statement to Senate Foreign Relations Subcommittee on Arms Control, International Law and Organization, in *Congressional Quarterly. Weekly Report,* vol. 28, no. 16 (April 17, 1970), p. 1019.

12. M. Leitenberg, *SIPRI Yearbook of World Armament and Disarmament 1968–1969,* Humanities Press, New York, 1969, chapter 2.

13. Galbraith, Ref. 7, pp. 237–240; also chapter 29.

14. See, for example, the discussion in *Science and Technology: Tools for Progress* (The Report of the President's Task Force on Science Policy, April 1970), pp. 38–39.

15. Quoted by J. Duscha in *Arms, Money and Politics,* Ives Washburn, New York, 1965, p. 2.

Summary of Discussion

Basic Research. It is very difficult to estimate accurately the financial contribution made to basic research in the United States in the interests of military applications. (See, for example, Curve A, Figure 4, p. 287 in the paper by F. A. Long). Estimates derived only from the defense budget tend to be too high when they include, for example, grants to universities to support pure mathematics which may not really be relevant to "military research." On the other hand there are federal agencies, such as the AEC and the NSF, that support basic research which does have possible military applications. The omitting of these funds tends to make an estimate based solely on defense appropriations too small.

National Security. The relationship between military expenditures and national security deserves much more attention than what it presently receives. The military commonly assumes that national security can be "bought" through expenditures on hardware and weapon systems. In this view a "high cost" budget implies "low risks" in terms of national security and a "low cost" budget implies "high risks." This assumption should be strongly questioned. Military expenditures form only one component of national security and it may not be the most important or critical component. Despite very large expenditures since World War II, the overall security of both superpowers has been gradually decreasing. There is an ample supply of historical miscalculations which were based on simply counting numbers of airplanes and guns. The relative success of the North Vietnamese and the Viet Cong, with comparatively unsophisticated weaponry, is a case in point.

There is a clear need for a broader point of view on national security that can adequately assess both the role of technology and hardware, and the many important political and psychological factors involved.

Aborted Military Systems J. P. Ruina

Both in the United States and the Soviet Union a very large number of major strategic military systems have been developed that were never deployed or that were terminated before deployment was complete. The abortion rate is high. But the even higher birth rate, combined with a low death rate for those systems that are deployed, is still causing an alarming increase in the nuclear-arms population. Examining the histories of aborted systems, we find that in some cases unforeseen technical difficulties were the prime cause for termination; in other cases, new technical developments caused obsolescence before completion; and in still other cases the conception was poor or the military benefit was inadequately analyzed initially.

I am referring here to the development of strategic military systems designed and intended for deployment and operational use, where expenditures are large (hundreds of millions of dollars at least), and where the primary technical effort is to satisfy military operational requirements. Such efforts may contribute to advancing the technical state of the art, but that is not a major intention. I include in the category of *research*, on the other hand, those exploratory developments where the original intention is to determine the technical feasibility of new

Jack Ruina was formerly Director of the Advanced Research Projects Agency, U.S. Department of Defense and President of the Institute for Defense Analysis. He is now Professor of Electrical Engineering at M.I.T. and also a member of the General Advisory Committee of the United States Arms Control and Disarmament Agency.

techniques or inventions. Often the demarcation between these categories is not very sharply drawn; but nevertheless my concern here is with large weapon-systems development and not with exploratory development.

A high rate of false starts which are carried far toward, or even into, the deployment stage does not seem to be an inevitable product of technological innovation. In other areas of national life involving technological products—communications, ground and air transportation, etc.—the fraction of new, major developments that are aborted is apparently much smaller than in the case of military systems. The defense business has had a plethora of Edsels. To be sure, one important contributing factor to the difference is the higher pressure for new technology in military hardware. But this is only one of several important ways that the world of military systems development differs from the world of civilian systems development.

As in the human analog, an ill-fated system once conceived is better aborted early than late. But any conception and development, even if later aborted, is more than a costly nuisance—it frequently has harmful long-term consequences. A false start may delay or inhibit the development of a more sensible solution to the same problem, if indeed a sensible solution exists. Also, military systems development activity in itself adds to the momentum of the arms race and to the national preoccupation with nuclear arms. In addition, any new systems development may provoke an adversary to create his own military system in response, and the response may take on a life of its own even if the original system is curtailed. Some military analysts have concluded that the Soviet Union's Tallinn air-defense system was designed to counter the U.S. B-70. The B-70 program was later cancelled but the Tallinn system deployment continued. The U.S. multiple independent reentry vehicle (MIRV) program was approved originally as a counter to the Soviet ABM system. But the U.S. MIRV program continues with new justification when Soviet ABM development seems to have diminished.

The Genesis of Weapon Systems

Military systems in our times are the product of an alliance of (1) political forces which are concerned with national security and protection, and (2) a dynamic and innovative industrial and technical community. We should understand why these partners are so deeply enamored of each other and live so well together, and what factors may contribute to the many ill-fated offspring of the union.

It is not hard to understand why political leaders and the public hold military technology and expertise in such awe. In the last few decades, technical developments such as the A-bomb, the H-bomb, radar, jet aircraft, and ICBMs have revolutionized strategic concepts. To the public each of these developments seemed spontaneous, the result of a technical breakthrough. It also seemed that if any of these had been created first by an adversary, we would have been at his mercy almost immediately. If the A-bomb and H-bomb were made possible by scientific discoveries and could be developed in great secrecy, why should the public not believe that other nuclear weapons could be developed in the same way with equal or greater impact? Why shouldn't the public fear the development of radiation weapons or new ballistic missile defense systems that they believe would give the adversary quick and definite strategic superiority? Why won't a new supersonic bomber or a manned space system have the same effect that the ICBM would have had if it had been fully deployed and only in the hands of one power? All of these technological possibilities seem equally impressive, equally important, and clearly within the capability of those nations that have developed the A-bombs, the H-bombs, satellites, and ICBMs to start with. I believe, for example, that public and congressional support for high-energy particle physics stems from the conviction that since past discoveries in this field had momentous military significance, future discoveries might have equal significance. The public doesn't really believe it when physicists tell them that research in particle physics presents little likelihood of discoveries of substantial military significance and that such research should be supported for its inherent scientific merit.

Until very recently we saw no harm in closing all perceived gaps in military technology. On the other hand, allowing any gap to develop seemed to be disastrous. Consequently, prudent and responsible national leaders would work hard at gap filling, whether the gaps were actual or potential.

An examination of the history of the Nike-Zeus ABM system is illustrative of these attitudes. The enthusiasts for the system predicted dire consequences if production and deployment did not proceed as quickly as possible. Conservative and cautious opinion was that the system might or might not have serious technical weaknesses, but until we were absolutely convinced of its shortcomings we must proceed with its production and deployment. A somewhat more daring view was that we could afford to delay the deployment decision but only for one year at a time. Almost nobody other than a small group of scientists in government (i.e., Drs. Kistiakowsky, Wiesner, and York and their advisors) was comfortable accepting, unequivocally, the view that protecting our people from Soviet missiles was impossible given the current technology and that deployment could be postponed indefinitely. They nevertheless felt that the country needed the *insurance* provided by a full-scale development and testing program for Nike-Zeus. The climate of the times was such that despite some private doubts, nobody openly advocated cancelling further development. Also interesting was that in the early decisions on Nike-Zeus deployment, escalatory effects on the arms race—action-reaction cycles that might be triggered—were not considered; the issue was strictly a matter of whether Nike-Zeus provided anything of direct military value for the vast expenditure involved. It was because objective technical analyses demonstrated very convincingly that it did not (and also because the system lacked the support of all the military services) that the production decision was delayed from year to year. But full-scale development and testing proceeded, at a total cost of over 2 billion dollars, until technology overtook Nike-Zeus and a new system, Nike-X, was designed for ballistic-missile defense.

But in the determination to avoid gaps, we created many military systems and devices which later clearly proved unnecessary, inadequate, or both. In the United States we expended billions of dollars on such developments as the Navajo missile, Dyna-soar, Midas, the Manned Orbital Laboratory, the B-70, Skybolt, nuclear-powered aircraft, the Sugar Grove antenna, and many more before they were cancelled. All of these systems were conceived and judged essential to avoid an existing or developing military gap but were later judged too wasteful to continue. It is not only the dollar waste on these systems that is to be regretted but also the waste resulting from the unnecessary noise and ruckus these programs added to the overreacting environment of strategic-weapons development.

The Soviet Union too seems to have its share of waste in military developments. We are, of course, less knowledgeable about the details of Soviet military developments, but to many Western military observers the Leningrad ABM of the early 1960s, the 100-megaton bomb development, fractional orbital bombing system (FOBS) development, the massive air-defense deployment within the Soviet Union, and the current Moscow ABM system are all examples of military waste in the broad sense referred to above.

Why the High Birth Rate

Now let me try to develop a very general framework for seeing why so much wasted and harmful effort is expended designing and developing military systems which are terminated before completion. A variety of mathematical models could be generated to describe the decision-making process involved in commitment to major military hardware developments, but for a gross qualitative description, three important variables should be included.

The first of these is the national tolerance for errors of omission—how willing are we to tolerate not having a military system which may later prove to be required? If we can judge from the past two decades, this tolerance is indeed low. In fact, in the strategic area its hard to find examples where we failed to press the development of a weapon system that was later missed and judged to be very important to the strategic balance.

Second, there is the tolerance for errors of commission—how concerned are we about starting on military programs which may later prove to be wasteful? Judging from the past again, the tolerance for errors of commission is very high. Although when a system was cancelled after a vast expenditure, Congress and the press frequently orated about military waste, in no way did this reaction compare to the panic when a strategic military gap—real or imagined—was exposed.

The third important variable is one that describes the inherent distinction, *ab initio*, between weapon systems that will "make it" and those that will not. Our ability to distinguish between these classes of systems, our ability to avoid errors of either omission or commission, is in good part subjective, and depends on our ability to assess all the technical, military, and political factors involved. But even with infinite wisdom, uncertainties inherent in factors entering into decisions cause a finite likelihood of error. First, the adversary's military and technical capabilities and intentions are never completely knowable. Second, in designing new military systems, the desire to avoid early obsolescence is so strong that we are always working at the brink of technical knowledge; we can never be sure whether we may be going over the precipice in costs or know-how. In addition, these uncertainties are exacerbated by the unavoidable time lag between conception and deployment, ranging from five to more than ten years, during which technological and political developments change greatly. Prognostication can never be perfect.

In testifying to Congress in support of a new bomber in 1968, Dr. Harold Brown, then Secretary of the Air Force, was very explicit about the inherent uncertainties he faced when making a decision when he said:

My reason for wanting an advanced bomber when we build one, to have the capability to fly supersonic at high altitude as well as being able to fly subsonic or barely sonic at low altitude, is that if you build a bomber of that sort, it is going to be around well through the 1980s, and *I am not smart enough to be able to anticipate what the best way to penetrate is going to be at that time.* (emphasis added)

In this statement Dr. Brown also indicates implicitly that on the matter of supersonic speed, he is more ready to make an error of commission than one of omission.

Reducing the Birth Rate

An examination of our three variables suggests several steps that could be taken to reduce the resources expended on, and the harm engendered by, "to-be-aborted" military systems and technology.

First, we can raise the threshold of merit and need that a proposed weapon system must demonstrate before it is approved for full-scale development. This would in principle raise the likelihood that we would have errors of omission but at the same time decrease the likelihood of having errors of commission.

Most objective observers, looking backward, will agree that realistic assessments of the strategic balance should have allowed both superpowers to raise this threshold for new systems development with very little danger to the military security of either nation. We see systems that could have been omitted, delayed, or acquired in smaller quantities without endangering the strategic balance; but we do not seem to lack systems we now need. This suggests that it is safe to start fewer major developments and to purchase less than we have in the past, thereby lowering the likelihood of waste and not necessarily risking the possibility of serious military "gaps." Reviewing the history of the last two decades should make us aware of unnecessary fears. Those of us who have followed strategic technological issues can cite repeated instances of overreaction to the threat to U.S. security from specific technical developments. Herbert York in his new book *Race to Oblivion* provides vivid examples of overreaction. He quotes a 1958 *Aviation Week* editorial written to support the U.S. nuclear-powered aircraft program:

On page 28 of this issue we are publishing the first account of a Soviet nuclear powered bomber prototype along with engineering sketches in as much detail as available data permits. Appearance of this nuclear powered military prototype comes as a sickening shock to the many dedicated U.S. Air Force and Naval aviation officers, Atomic Energy Commission technicians, and industry engineers who have been working doggedly on our own nuclear aircraft propulsion program despite financial starvations, scientific scoffing, and top level indifference, for once again the Soviets have beaten us needlessly to a significant technical punch.

Now, twelve years later, neither we nor the Soviet Union have yet overcome the technical difficulties in developing a usable nuclear powered aircraft.

York also refers to the fears, instilled by the Sputnik launching, that our military space program was lagging dangerously and quotes an industrial leader as saying, "Tomorrow the country that controls the moon will control the earth."

In the early 1960s there was fear that a comprehensive test ban treaty would create the conditions that would allow the Soviet Union but not the United States to develop a neutron bomb. Senator Dodd expressed this concern when he said, "If the Soviets possess the neutron bomb in quantity, its far greater effectiveness, its relative cheapness, and its freedom from widespread devastation and from fallout, would give them an advantage of critical proportions."

Dr. Teller was apprehensive that the limited

test ban treaty would freeze our lag in knowledge about nuclear effects. In testifying to a Senate Committee in 1963, he said:

In 1960, he [Khrushchev] wasn't willing to sign, but now he had these magnificient test series of 1961 and 1962. He now knows how to defend himself. He now knows, probably, where the weaknesses lie in our defense. He has the knowledge, and he is now willing to stop and prevent us from obtaining similar knowledge.

Fears of imminent gaps must also have existed in the Soviet Union but unfortunately we have little access to relevant Soviet documents.

We have not only been overly fearful that a gap would develop in one or another of the many dimensions of the arms race, but also we have often exaggerated the consequences to our strategic posture if a real gap should occur in one of these dimensions. How serious would it really have been if the Soviets had indeed succeeded in developing a nuclear-powered aircraft or a neutron bomb and we had not? The strategic nuclear balance has not really been so fragile that every occasional gap would upset it. Deterrence capability is not weighed on a fine scale. That is not to say that every gap was tolerable; but perhaps more were than we have been willing to admit.

We might even speculate whether our deterrent posture vis-à-vis the Soviet Union would really have been much different if we were without one of our very "essential" systems (Polaris, our bomber force since Polaris, Atlas, Titan, Minuteman, etc.), I am thinking here of the effect on our deterrence capability in the direct military sense. I realize that the size of our nuclear forces may have important implications for our diplomacy. For example, there is speculation whether our nuclear superiority was a factor in causing the Soviet missile withdrawal from Cuba—to the extent that it was a factor, it was not a consequence of any inadequacy of the Soviet deterrent capability but rather a response to psychological effects which I am not addressing.

All this suggests that it is safe for us to raise our justification requirements before initiating new weapon systems. Raising the approval "threshold" would result in reducing the number of false starts and would not seriously risk the development of any intolerable military gaps.

In the past year we have seen change in U.S. attitudes to military systems. Budgetary pressures are causing less tolerance for false starts in weapon-systems development. The public as a whole is also becoming increasingly sensitive to the sequence of escalating effects that the introduction of new systems may trigger. The fear of arms escalation is one of the important restraints now in U.S. ABM deployment; yet only one decade ago it wasn't even considered when the Nike-Zeus deployment decision was being made. Now national leaders including the president are expected to discuss escalatory effects publicly when justifying a new weapon development.

There is also increasing public belief in the great stability of the strategic balance, and with it, a growing conviction that significant gaps in strategic arms are apparitions, visible only to those who partake in certain peculiar nuclear numerology. We can observe, for example, how relatively calm the U.S. public is, in spite of the exhortations of some of our high defense officials regarding the massive and mysterious buildup of the Soviet nuclear missile force. We can all speculate about the kind of reaction that a similar development would have provoked five to ten years ago.

Decreasing Uncertainty

Minimizing unnecessary military systems developments also requires minimizing the uncertainty we must deal with in our decision-making for weapons development. Maintaining the strategic balance with a minimum of futile effort requires knowledge about what is in each of the two scalepans, what is likely to be added, and what could conceivably be added. Without knowledge, each side feels coerced—indeed, panicked—into adding all he can of weighty substance to his inventory. The Soviet Union, for reasons of tradition and narrow concepts of military security, has been secretive about its military posture, its strategic policy, and its plans and programs. In contrast, the United States for reasons of policy, tradition, and ineptness in keeping secrets, has been so open that Soviet military planners perhaps know as much about these matters in the United States as U.S. military planners. All too frequently our gaps in knowledge about Soviet capabilities have fueled unnecessary fear and distrust. I am convinced that if the amount of information the United States has about Soviet nuclear forces were to increase, the creation of new military nuclear armaments would decrease for the United States and in turn for the Soviet Union; conversely, that if the amount of information we now have about the Soviet Union should decrease in any substantial amount, it would provoke a tremendous number of new military developments on both sides.

An assured high level of information flow is essential if we are to achieve a low level for our deterrent forces, reduce our fears, and minimize our efforts in what John Foster, the Director of Defense Research and Engineering, describes as "[making] sure that whatever [the Soviets] do of the possible things that we imagine they might do, we will be prepared."

In urging greater effort at improving information flow I exclude, of course, such information as would increase our adversary's first-strike capabilities. Technical details about penetration systems that would help an adversary's defense, location of missile-carrying submarines, etc., are best kept secret.

Whenever technical developments have permitted an increase in information flow, the relaxing effect on U.S. military programs was felt at all levels of government. Chalmers Roberts in his book *The Nuclear Years* quotes President Johnson as having said in 1967:

We've spent $35 or $40 billion on the space program. And if nothing else had come out of it except the knowledge we've gained from space photography, it would be worth ten times what the whole program has cost. Because, tonight, we know how many

missiles the enemy has. And, it turns out, our guesses were way off. We were doing things we didn't need to do. We were building things we didn't need to build. We were harboring fears we didn't need to harbor.

Arms Control and Uncertainty

Within the framework I have used in this paper, it is interesting to examine the effects of a United States–Soviet agreement limiting strategic arms. Deploying a system prohibited by agreement would mean either violating or abrogating the agreement and would undoubtedly require far more internal justification than the same action without the agreement. Therefore the likelihood of an error of omission is also higher if it is assumed that an adversary may not be restrained by a treaty, or if it seems necessary to develop a prohibited response to the adversary's actions in an unprohibited area. To compensate for the added risk of having a gap and to raise the other party's threshold for initiating a prohibited program, the United States has always desired verification procedures that would add to the information it would have without an agreement. This verification requirement has usually been in the form of on-site inspection, but "black boxes" have been suggested as alternative information sources. The Soviet Union has considered this information collection as an intelligence activity (of a highly restricted sort, of course). Both designations have some validity in that the aim is for more information. However, verification connotes a legitimate activity while intelligence connotes a nefarious activity. How much additional information and what new channels are necessary to compensate for the restraint on action that a treaty imposes depend on the confidence we have that the treaty will not be violated and on what information channels already exist.

Maintaining Research

Also important in reducing uncertainty in military systems planning is the maintenance of an intensive program of basic and applied research (in contrast to full engineering systems development for operational use) and technical and political analysis.

In the last decade both superpowers have had intensive military research programs. I expect every actual and potential area of interest in military technology has been or is being fully explored in both countries. I believe that with the intensity of our ongoing research, we are less fearful of an opponent's scientific discovery of the technological breakthrough that would quickly and decisively change the strategic balance. That is not to say that one country or the other might not be "ahead" in some area of technology, that one country might not be better than the other in developing chemical propellants, or ballistic-missile guidance systems, or reentry vehicles, or high-power radars. But a lead, in time or capability, has very little likelihood of substantially changing the strategic balance.

For example, we feel secure in our knowledge of ballistic-missile defense because we have spent many billions in research in this area. Without the intensive U.S. R&D program in ABM research we would be extremely fearful of Soviet ABM developments, Soviet ABM tests, and Soviet ABM deployment. Dr. Teller's fears about what the Soviets learned from their test series lost much of their credibility when our political leaders and our scientists reviewed what we had learned from the large number of U.S. nuclear explosions. A great deal of effort went into ballistic-missile defense, and our findings demonstrated only the remote possibility that a few Soviet explosions could have given them so much information as to provide them with a clear and substantial advantage.

I argue, therefore, that an intensive research program can be an important factor in allaying fears and in reducing military systems activity in both countries.

I am mindful of the fact that military research presents our military and political leaders with alternatives which they find hard to resist, and of the consequent argument that the only way to stop new military systems development is to stop military research. Some writers refer to a "technological imperative" at work—that is, if a weapon can be made it will be made. There is no doubt some truth to this, but the concept is overly simplistic. There are restraints to the temptation to develop and deploy, without discrimination, the technologically possible. These restraints can and must be strengthened. We can derive some hope from the past when we did restrain ourselves from developing shipborne nuclear ballistic missiles, bombs in orbit, 100-megaton bombs, and many other technically feasible systems. Now we must extend this restraint to cover (rather than covet) a larger class of the technologically possible.

Besides, if we did attempt to limit research, there are some very practical difficulties involved. Is it realistic to expect one side to limit its activities and not the other? Can either side ever be confident that the other is not learning his R&D? Can military research really be separated from civilian research? Or should we propose a limit on *all* research? I must repeat that I speak here of research and exploratory development. Full-scale systems development is of course a different matter.

In summary, my personal view is that a high level of activity in military research and analysis has benefits that exceed its overall costs (in dollars and to the arms race). On the other hand, maintaining the large number of full-scale weapon-system developments we have had is unnecessary and harmful, and some reduction is in order with or without an agreement. This still leaves a large amount of technical activity, mainly in the testing area, somewhere in the middle, warranting careful examination. My own predilection is to try to limit, but not eliminate, full-scale testing by agreement. There are, for example, clear (although not pressing) advantages to limiting underground testing to small yields (less than 10 kilotons), and an agreement to do this should not have to face serious technical or political obstacles. There are also good reasons to try to control the kind and number of full-range missile tests by agreement.

Comparison with Civilian Programs

In describing the ambience for weapon-systems development, it is hard to resist comparison with attitudes toward the use of technology for national domestic needs. First, there is amazing tolerance for errors of omission in the civilian sector and a very low tolerance for errors of commission. The approval threshold for new systems is set far higher than in the military sector; waste is unacceptable, whatever the possible gain. Second, uncertainty about how to cope with the "threat" doesn't arouse fears which stimulate action, but rather dulls us into inaction; and yet the threat to our well-being and our security can be as real and as vivid as the military threat.

If the approval threshold for new civilian systems had been set as low as it has been for military systems, we might have developed, tested, and deployed new marvels of technology to dispose of our waste products, to handle urban transportation, and to provide health services for all. Many systems in the course of development would not prove to be economically, technically, or operationally feasible and would have to be aborted; but the flow of new systems would continue with a mad momentum we would learn to accept.

The Role of Experts

I would like to deal briefly with the role of the scientific and technical expert in the weapon-systems decision-making process.

Political leaders quite understandably are generally not sophisticated in their understanding of science and technology and they have to call upon experts to help in assessing risks and uncertainties in areas of new technology. Experts with impeccable credentials come perhaps not in 28 delicious flavors, but in a sizable array of professional background, shades of opinion, degree of optimism, political sensitivity, and readiness to participate in the political process. Because of this variety, a political leader does not have to search very hard to locate an expert whom he finds *simpatico*. Secretary Laird would not be likely to choose Linus Pauling to counsel him nor would Senator Fulbright clutch Edward Teller to his congressional bosom. Indeed, we must worry about how to make sure that national decisions are derived from consideration of a variety of perspectives. The decision-making process should include inputs from different perspectives to provide the "checks and balances" on technical judgments that complicated decisions require. Although it may be only human to look for our own reflections in other people, it would be exceedingly detrimental to our national policy to shut off the broad spectrum of viewpoints that exist in the scientific and technical community.

Occasionally, the expert's role in a decision is clearly delineated, as was the case in the Cuban missile crisis. Expert photographic interpretation made it possible to determine the nature of the Soviet missile buildup; but it was clear that this could only provide the basis for the political decisions that were not the province of the technical experts.

More often, however, it is difficult to distinguish between what is technically known and already understood, what requires technical prognostication and is therefore more judgmental, and what is clearly political. Indeed, most decisions involve a fabric of subtly interwoven threads of each of these factors so that isolation of any one is impossible. Because of this, experts themselves cannot separate the woof of their technical knowledge from the warp of their general perspectives. And, therefore, I emphasize again how important it is to include varied expert inputs in decision-making.

The possible variations in expert opinion, combined with the inherent complications of the bases for decisions, create some serious dangers in the decision-making process. Because I know it so well and because, in retrospect, the problems seem clear, I shall illustrate by citing some of the happenings in the nuclear test ban debate.

First, I would like to mention the danger that something may be considered a technical question when in fact it is not. During the nuclear test ban debate, a major issue was the precise number of annual on-site inspections necessary for adequate verification. The United States numbers varied from 7 to 20 and the Soviet numbers varied from 0 to 3. It was commonly accepted that there were profound technical reasons associated with the number of inspections and that our security was clearly in jeopardy with 3 or 4 or 5 or 6 inspections per year, but not with 7. I suppose the Soviets also believed there was some fundamental technical basis for their limit of 3 to the number of inspections per year. Now we can look back in amusement or dismay at the presumed merits of this nontechnical technicality that prevented any agreement on underground nuclear testing.

Another major problem in the decision-making process can come from rationalizing our basic concerns by an almost total preoccupation with narrow technical issues when in fact we are confusing the part with the whole. In the course of the nuclear test ban considerations, the state of seismology became a major issue. Was it going to be possible to distinguish seismic signals of earthquakes from those of underground tests? Could our adversary conceal tests? How small a signal could be detected? Incidentally, Dr. Carl Romney in congressional testimony said that he believed tests of three kilotons were the lowest yield detectable, and I said I believed tests of one kiloton. In fact, we did not disagree at all; but the public did not know that seismology is sometimes described as a "factor of two" science and attributed great significance to the difference in these numbers. But what is most important to point out is that everyone was so involved with these fine points that the big issue was lost. We should have been asking how much it really mattered if advances in seismology improved our detection and identification capability. We knew that it was always going to be almost impossible to detect very low yield tests by seismic means. We could already detect and identify very large tests. The emphasis should have been on the effect of tests in the uncertain middle region on the strategic balance. I would argue that all of the complexities of the technical arguments in seismology made us think that we were asking the right questions. Here, I think the experts themselves helped muddy the issues.

In summary then, we have to recognize that because technological expertise is valued, feared, and sometimes intimidating, we risk falling into a technological trap from which extrication is difficult.

It is indeed a hopeful sign that it is the scientists and technologists themselves who are sounding the alarm that they are taken too seriously and too completely, and that major weapons decisions are too important to be left to the "experts" and must be weighed in the larger political sphere.

Summary of Discussion

The Genesis of Weapon Systems. The suggestion was made that the underlying force behind the many systems that either reach deployment or are aborted is the interaction of the fear and misunderstanding of technology on the part of political leaders with the support from politically compatible technical experts. The political leaders get trapped by the technical details of the systems and thereby lose sight of the harder, nonquantitative, but more important overall security questions. The ease with which funds are made available is a symptom of this cooperation between the politicians, who provide the money, and the technologists, who use it. It was also pointed out that the intelligence community must be included in this analysis of the dynamics of the birth of military systems.

Furthermore, internal as well as external pressures may also play an important role in the creation of new or improved systems. Although the American development of the MIRV may have been partially a response to the Soviet ABM deployment, it was also, in large measure, a response to American ABM research and to the pressures to carry to fruition ideas which seemed technically interesting.

Implications for the Future. Today, with sophisticated strategic systems, great destruction can be caused easily and cheaply. There is also a variety of alternative types of systems which can be employed to guarantee this capability. As a result of these facts, a sudden imbalance is unlikely to develop in the foreseeable future. Therefore, the threshold for omissions, or at least delay, can be safely lowered in order to raise the threshold against creating new systems. The more worrisome the risk of omission, the more likely it is that new systems will be started and the less likely that a treaty to limit new weapons will be achieved.

Until recently it was widely believed that starting a new system involved no great risk, and decisions about whether to go ahead with new developments were made largely on the grounds of the availability of resources. Today there is an increasing realization that starting a new system itself represents a risk of escalation because of the dynamics of the arms race.

Comments and Summary of Discussion on Military Research and Development

Benign Technology: Comments by R. L. Sproull

Not all technology is pernicious. Although we are here quite properly trying to identify, and if possible to devise ways to control, major new technological developments that might be destabilizing, it might be well to consider briefly two examples of possibly benign technologies.

The first is any technology that would make the doctrine of "launch on warning" less attractive. We can safely assume, I believe, that techniques of providing warning of a ballistic missile attack will be continuously developed. At some stage of this development, confidence in the warning mechanism (or combination of mechanisms) may grow to the point that a doctrine of launch upon warning would seem preferable to a doctrine of absorbing the attack before retaliating. Yet the endless possibilities of mistakes, electronic goofs, and *n*th country mischief render such a doctrine intrinsically dangerous. Given enough time, *some* combination of accidents will eventually defeat the electronic safeguards, the "hot line," and common sense.

Alternatives to such a doctrine ought to be developed, and my main point is that incentives ought to be created so that such development is actively pursued. Such incentives might be incorporated into an arms control agreement or might simply be provided by each nation on its own.

Let me give an example of such a development but without any claim that this is the best approach. A warning that a small number of threatening objects was approaching might produce as a response a mid-course intercept. The usual arguments against mid-course interception would not apply since the number of objects was small. Although this move would not be an acceptable counter to a warning of a large salvo, most false warnings would presumably indicate only a small number of objects. The development of a mid-course, nondiscriminating interception capability might thus be a healthy action, providing a nation "something to do" in response to a warning other than an almost automatic retaliation.

The technology of verification of arms control agreements is another example of benign technology, especially familiar to students of the partial nuclear test ban. An interesting possibility of a future agreement is a ban on all flight faster than the speed of sound. The technology already exists to verify such an agreement except possibly in some areas, where remoteness or large traffic is a problem, and there would appear to be no fundamental obstacles to developing the technology for comprehensive verification.

Development of new supersonic aircraft would presumably cease upon the adoption of this agreement. Lack of training and qualification flights would presumably destroy the usefulness of existing supersonic aircraft, and therefore they would soon disappear from force structures.

Such an agreement should therefore have substantial advantages for defense budgets in both large and small countries. It might be a necessary partner to any agreement limiting quantities of aircraft, and it might substantially facilitate distinguishing air-defense from ballistic-missile-defense installations. (Verified exceptions would probably have to be made for space launches and reentries and for training-launches of missiles.) Obviously such an agreement raises many problems, but if substantial progress is to be made in reducing defense budgets, schemes as ambitious as this will be needed, and technological development will be necessary to make the agreements verifiable and acceptable.

Lessening of Tension Coming from the Fear of Secret Scientific Advances: Comments by J. Guéron

Three ways were pointed out to counteract the risk of secret scientific advances: citizen control; budget control; and international scientific cooperation.

In most countries, including the Western European ones, citizen control, as understood in the United States, hardly exists. This way of political expression is not a traditional one, nor is it favored by the legal and constitutional situation.

Budget control by parliamentary action becomes therefore all the more important. Here again, the

U.S. system of hearings by congressional committees is rather unique. Elsewhere the connection between government officials and informed citizens is very weak, unless the latter get involved through their private political, not professional, capacity.

Thus, subjects such as those we are considering escape the attention of the citizen, unless they become significant enough to appear as a major issue in a national election (which has never happened so far).

Efforts to educate the public by groups such as the Pugwash Committees are therefore all the more important.

Active international scientific cooperation reduces the risk of unilateral technical breakthroughs resulting from secret discoveries followed up by swift developments. However, one must emphasize that a few limited cases of such cooperation between countries from the East and West—however valuable they may be as demonstration and initiation cases—cannot provide an effective lessening of tension.

A very extensive and widespread continuous cooperation would be necessary. As often shown by sad examples from Western Europe (see my papers in the *Bulletin of the Atomic Scientists*, October 1969, and June 1970), scientific cooperation in fields leading to applications of science can endure and become politically and economically effective only if it develops to the point of a joint sectorial R&D policy, backed by the proper patent practice, industrial structure, and government purchasing system.

Restricting Research and Development: Comments by J. Prawitz

It is an old question whether it is possible and feasible to put restrictions on research and development activities in order to prohibit possible expansion of the arms race into new areas. It will not be news to this audience that it is also a very difficult one. It is indeed impossible to deal with it in a few lines, and I shall restrict myself to a few comments only.

The central difficulty is of course that a piece of research can serve the purposes both of peace and of war. In the latter case it can be good for both offense and defense. One example is microbiological research promoting at the same time both public health and the development of biological warfare or protection against such warfare.

Another problem is that control measures, established to verify international agreements, must be easily described in treaty terms and they must be easily formalized to become suitable as a basis for political decision-making and action.

There is, therefore, no general way of controlling military R&D, as this control would sometimes have to deal with the intentions and thoughts of individual scientists. In specific areas, however, practical measures might be envisaged.

One example is R&D activities involving operations of such a magnitude that distinct features of obvious significance for weapons development can be easily identified and observed in a formalized manner. The obvious example, of course, is the nuclear test ban, prohibiting activities where one single experiment has dimensions of geophysical size. On the same grounds, a MIRV-test ban has also been suggested.

The question, then, is whether there are attractive indirect ways to restrict military R&D in order to stop potential new arms races in new areas of potential confrontations. Here, certain recent arms-limitation agreements are of great importance. The Antarctic treaty, the outer-space treaty, the non-proliferation treaty, the sea-bed treaty about to emerge, etc., will be of great significance. It is also worth mentioning the further restrictions on biological and chemical weapons now under serious negotiation. These have been criticized because they do not result in real disarmament. This may be true; but the fact that they exclude militarily untouched areas from being flooded by spectacular new weapons is important enough to justify them and to invoke further measures of this kind. In particular, it would be desirable to conclude such arms-limitation agreements now, before these areas have been massively invaded by either military or peaceful activities, *in order to establish first the security framework within which peaceful activities can then develop.*

This would exclude such new sectors from the arms race, make intended military R&D meaningless, and justify peaceful developments only.

Another indirect type of restriction would be the internationalization of research efforts in the new areas. This does not necessarily mean the setting up of new international bodies to take over activities now national but rather to fulfill the function of opening up research results for common observation.

To make this effective, the international cooperation must be extensive enough to make it definitely attractive for institutions to participate actively. The efforts now undertaken by the United Nations to establish an international cooperation scheme are of great significance for this. Examples of possible future significance are weather and climate manipulations.

A last comment touches on the area between R&D and production. In certain cases the size of research operations overlaps that of small production and deployment. This is true, e.g., for bacteriological laboratories, from which a biological attack might be launched directly out of the peacetime research operation. Control, aiming at early detection of possible attacks of this kind, could be limited to an overall assessment of the nature and size of what is going on in the various laboratories. Studies on this problem are going on right now in various places in the world. Other examples of this kind may be found to arise in the future.

Summary of Discussion

Importance of the Problem. We are faced today with several new weapon systems, ABM and MIRV, which are at the stage of being deployed. The question to which we addressed ourselves is: Could the development of these systems have been avoided, and can future weapon systems be avoided, by some sort of national or international policy on R&D?

We see today that technology which was considered remote or even unachievable only ten years ago has in fact been realized. Startling breakthroughs in computer technology and missile guidance are now taken for granted. Indeed, in retrospect, these advancements were made far more easily than could have been imagined.

It seems that there is a general principle operating here: whatever appears to be even remotely possible turns out to be easy. This observation has frightening consequences, for scientists and engineers seem to have accepted the challenge of constructing whatever is possible. It is difficult to be optimistic about the prospects of changing this policy in the future.

Research and Development and Destabilization. There are two opposing viewpoints on the role of R&D in the arms race. One holds that a vigorous, broad program of research is necessary to avoid technological surprises by giving each side the basic information on what is technically possible. Without this information worst-case analysis (WCA) will always dominate the decision-making. Furthermore, since it is operational development which gives a true measure of one's intentions, research, in itself, is not destabilizing. Moreover if there are weapon systems—such as nuclear submarines—which have a clearly stabilizing effect, then further R&D to improve these systems is certainly desirable.

The other point of view emphasized the destabilizing effects of research and especially development. This viewpoint holds that the U.S.–USSR strategic arms race is only partly based on political or territorial controversy and to a substantial extent is simply self-perpetuating, i.e., it feeds on itself. Continuing R&D, generated by the arms race itself, is likely to lead to further escalation and destabilization. The mechanism for this is straightforward: the purpose of a system under development is frequently ambiguous, especially in its early stages. This uncertainty leads to a worst-case analysis by the other side, which in turn precipitates a complete spectrum of responses to cover all possibilities. Many new programs and superfluous countermeasures result.

Controlling Research and Development. There is an apparent dilemma facing those who would restrict the R&D of weapon systems. One would ideally like to halt the development of systems before they have become too large and thereby attained a life of their own. Yet, because the early stages are often very inconspicuous and in many cases appear to foreshadow developments that are largely benign in nature, development can often not be stopped until it is quite advanced.

In the United States, the Department of Defense contributes about $200 million a year to basic research. It is especially difficult to control development within the Department of Defense itself because of the many small-scale projects that are continually being spawned. It is much easier to work, on an international level, at preventing testing and deployment of actual operational systems. For example, one could, by international cooperation and monitoring, prevent the operation of supersonic aircraft. On the other hand, from the point of view of the arms race, it is unlikely that the mere limiting of supersonic military aircraft would be sufficient. The military would simply maintain their older aircraft—perhaps at a greater overall cost.

It is clear that R&D restrictions must be an integral part of a larger program of arms limitation. Indeed, if only limits on the number of weapons are agreed upon—with no basic change in attitudes—it is likely that qualitative improvement of weapons will become the goal, with still more emphasis placed on R&D.

Scientists and Research and Development. Several possible schemes to discourage R&D in a given area of weapon research were suggested. One proposal would operate through legal prohibition. In principle one would publicly seek to make it illegal to work on or develop a given weapon system, such as a chemical or biological one. Such a prohibition would admittedly be difficult to define and even more difficult to monitor, but it is perhaps worth pursuing to see if it is at all feasible.

A somewhat analogous proposal would rely on the voluntary refusal by scientists, acting presumably as individuals, to work on a given weapon system. As an example, it was suggested that in the area of biological and chemical weapons the lack of any dramatic breakthroughs in developing such weapons could be attributed to the unwillingness of outstanding, first-rate biologists to involve themselves. This suggestion was countered with the explanation that physicists had already produced enough deterrence in nuclear weapons and there was therefore no pressing national need to develop biological or chemical weapons. Whether voluntary actions on the part of individuals could serve as a limitation mechanism on weapon research was left an open question.

Another, more indirect approach to limit weapon research would involve the initiation of research programs designed to attract bright, young scientists and engineers into new, challenging areas of nonmilitary research. Environmental pollution is one obvious field. One could hope for much more East-West cooperation in these areas.

Controlling Basic Research. Controlling research outside of the Department of Defense through the regulation of general, basic research and knowledge is unacceptable in our society. For one thing it is impossible to predict in advance where a military application will arise. Furthermore, military research is very similar to ordinary basic research. Because of this similarity it would be very difficult to generate only "benign" basic research.

Limitation Treaties and Prohibition Treaties. There is a distinction which should be borne in mind when one speaks of controlling the development and deployment of weapon systems. The distinction is between *limiting* the number of a given kind of weapon and *prohibiting* the weapon altogether. So far, our international agreements have been to prohibit. Thus, for example, we agreed not to place *any* nuclear weapons in orbit around the earth. Prohibiting a weapon system is relatively easy to achieve and monitor, and R&D would be expected to decrease sharply in an area where such an agreement was reached. Such, for example, is the case with chemical and biological weapons in the United States. Agreements to limit the number of weapons are much more difficult to monitor, particularly if the number of permitted weapons is small. If the number is large, deviation that might go undetected may not be considered significant. That is the present reality with respect to ICBMs in the United States and the Soviet Union. As stated earlier, there is a genuine fear that limiting agreements will simply shift the focus of activity to new R&D in an effort to improve the current systems in a qualitative way and to avoid the possibility of technological surprise. Unfortunately, such new qualitative improvements can be destabilizing. In negotiating treaties that limit the number of a given kind of weapon, one should also strive for qualitative restrictions to avoid such a destabilizing R&D race. It is also imperative to keep reviewing new technology and to keep any new developments international. In this way one can hope to introduce international restraints.

The Need for International Cooperation in Research. The question of whether continued, vigorous R&D on weapon systems is destabilizing or not depends for its answer on a host of technical and political considerations and there seemed to be no consensus reached on the issue during the discussion.

There was, however, almost unanimous agreement that more international cooperation in research would have a definite, inhibiting effect on the arms race by minimizing the possibility and the fear of surprising, unilateral, technological breakthroughs. The hope is that international control agreements will be more easily reached if neither side has a substantial lead in exploiting new discoveries for military purposes. The fear was expressed that, if a control agreement could not be reached, such cooperation would have little effect beyond insuring that both superpowers would start the next phase of the arms race on an equal footing.

Examples of cooperative basic research were cited in which U.S. and Soviet scientists were collaborating in oceanographic investigations. Many participants were most anxious to go beyond the stage of merely talking together to the point of working together. They also stressed the importance of scientists becoming involved in programs designed to achieve a lasting peace and expressed the conviction that, as scientists, they had a particular responsibility to bring their technical understanding of the arms race into the arena of political decision-making.

4 The Political Implications

Political Implications

I. M. H. Smart

After all that we have heard in the last two days about the prospective counters to deterrent offensive forces—the prospect of greatly improved accuracy for guidance systems, the possibility of effective, if very expensive, ABM systems, and the possibility of a transparent ocean for ASW purposes—it might seem that talking of the future of offensive systems was rather like talking of the future of dinosaurs. Fortunately, I suspect that dinosaurs may serve our purpose, provided that we recognize that purpose clearly.

If our purpose is to fight a nuclear war, there is no doubt that the potentially available countermeasures which I have mentioned are worrisome. If, however, our purpose is to deter the outbreak of a nuclear war, I would suggest that the grounds for concern are both slighter and less obvious. I do not mean that we should become complacent but only that we should stop short of panic.

For the last twenty-five years, we have spoken in terms of deterrence and have purported to rely upon it for our security. I suspect, however, that we have, in fact, continued to think and feel as though we remained in an old style offense-defense environment and that we have never really trusted the operation of deterrence to the extent which our words have suggested. Evidence in support of this thesis might be drawn from a number of sources. It seems to me, for example, that the policies with which we continue

Ian M. H. Smart is Assistant Director of the London-based Institute for Strategic Studies.

to protect and classify information about weapon systems offer some evidence of this sort. If we fully trusted the operation of deterrence, it would ostensibly be logical for us to take whatever steps were necessary to insure that the size, basic characteristics, and operational performance of our retaliatory forces were well known to any potential adversary. In fact, as we know, this is very far from being the case. Parallel evidence seems to emerge from what was taken to be the apparent implausibility of the "cooperative verification" suggested by Dr. Fubini. In present political circumstances, this does, indeed, seem to be an implausible step. Nevertheless, it is worth remembering that, if we genuinely relied upon deterrence for our security, it would also be an entirely logical and reasonable one. It may in fact be that, while we construct and deploy weapon systems for deterrent purposes in the hope of reinforcing our own security, we surround them with attitudes which tend to diminish their effectiveness for this purpose.

The implication of all this seems to me to be that we have never really faced up to the nature of deterrence. By the same token, I suspect that we have never faced up in recent years to the question of the minimum levels at which deterrence can be expected to operate. By this, I do not mean only that we have not considered adequately the minimum damage the threat of which may be sufficient to deter but also that we have failed to recognize the minimum probability of sustaining such damage which can be relevant to the same purpose. As a personal credo, I would state that, for all relevant purposes, the initiation of major nuclear war can probably be deterred by a much lower probability of inflicting much less damage than is now commonly acknowledged.

This assertion is directly relevant to the question of dinosaurs. Although a great deal of historical evidence suggests that we should be cautious about the effectiveness of countermeasures, prospective developments *may* create some theoretical ability to destroy the bulk of land-based missiles, to locate and destroy a large number of the submarines which carry submarine-launched ballistic missiles (SLBMs) for retaliatory purposes, and to destroy the bulk of strategic bomber aircraft on the ground or in the air. An extreme "worst-case calculation" might even suggest that prospective countermeasures could do all these things simultaneously and in a coordinated manner. In fact, however, even the most confirmed pessimist would have to concede that, in each of these cases, there will remain a margin of uncertainty —a marginal probability that the countermeasures will fail. The important fact is that, when the marginal uncertainties are multiplied together, the product tends to operate powerfully in the direction of deterrence. Even if one were to take the apparently ludicrous case of having a 95% probability of destroying 95% of the retaliatory forces on the other side, the multiplication of the reciprocal probabilities of failure would result in a residual level of retaliatory forces which might not, despite common belief, be entirely inadequate for many deterrent purposes.
I suggest, in fact, that, to take an extreme case, a 10% probability of receiving a retaliatory strike from something over fifty separate megaton-range weapons

would have a powerful deterrent effect in a wide range of circumstances. In fact, of course, when all the factors affecting reliability and performance of countermeasures are multiplied together, it is almost certain that one would be dealing with a much higher probability of significantly heavier retaliation than this.

To extend the "mine-field complex" of which Dr. Fubini has spoken, it is clear that any nation, like any individual, is likely to be deterred in almost all circumstances by a 99% probability of death. What I am suggesting is that, for very many purposes indeed, a nation or an individual would also be deterred by a 50% probability of losing all four limbs.

Our failure to think clearly about the operation of deterrence and about the minimum levels at which it may be expected to operate effectively marks, as I see it, a departure from the constraints of any external rationality. A parallel departure seems to have occurred in the case of the strategic-arms race itself. Here, as in the case of deterrence, the departure is encouraged by what we have been discussing as the process of "worst-case calculation." Professor Rathjens has spoken to us of two forms of instability: "arms race instability" and "crisis instability." I was delighted that Dr. Resnick added a third category to these: "analyst instability." Personally, I suspect that this third form of instability is the most worrisome of all. I had a strong feeling in this sense during our earlier discussion about reconnaissance and surveillance, and about the alarm created by the possibility of cheating systems for the verification of agreements. Some possibility of this type must always exist in the case of any verification system. Too frequently, however, our alarm about it arises from failure to distinguish between the purposes to which verification is relevant. For many purposes, such as the safeguarding of special fissionable material, we are bound to aim at what I would describe as "insurance verification." In other words, we will want verification measures which offer an extremely high probability of detecting any violation whatever. I see no evidence, however, that this approach is relevant to guarantees of deterrence, for which I would consider the name "deterrent verification" more appropriate. In other words, I would consider that, especially when one is dealing with a political agreement in which governments have vested a significant measure of political credit, the penalties—political and other—for cheating against an arrangement involving deterrence are so great that much lower probabilities of detection will suffice in the case of verification.

In all the areas which I have mentioned—levels of retaliatory damage, levels of probability of sustaining such damage, and impermeability of verification measure—it seems to me that we have been seriously handicapped by our consistent failure to analyze the general character and specific operation of deterrence as a phenomenon which is not only one of military force but also one of psychology, sociology, and politics. Perhaps unfortunately, this failure has not so far had serious implications for the procurement of strategic-weapon systems. During the last twenty-five years, the sorts of systems which we would have acquired for offense-defense purposes would not, in any case, have looked very different from those which we would have acquired for purposes of deterrence. The relative ease of living with an ambiguity of this sort may, however, be a luxury which we are about to lose. A number of the developments which technology now offers, including such types of weapons as extremely accurate offensive reentry vehicles and relatively effective area-defense ABM systems, seems to me to confront us with the choice which we have so far been able to avoid. In order words, they confront us, for the first time, with the need to choose between weapons for deterrence and weapons for offense-defense. In this respect, the day of the multiple-purpose system may, in many respects, be behind us.

We have always, of course, been handicapped, as we all recognize, by the phenomenon of worst-case calculation. Our primary error in this respect has been to represent extrapolations from current capability as though they were also predictions of future intention. We have become used to living with relatively good evidence of each other's capabilities in the strategic-weapons field. We continue, however, to base a large part of our judgments about each other's intentions on stereotypes rather than on direct evidence. In such circumstances as these, it is hardly surprising if many of the judgments which are expressed in terms of intentions are, in fact, founded upon evidence of capability. Capabilities, indeed, come to dominate almost completely our measure of reciprocal intentions. To the suspicions which are thus inevitably generated, there are added two other phenomena. The first is what seems to be the inexorable march of technology and the ineluctable temptations which it presents to nontechnical decision-makers. The second is the fact that, because of the lengthy periods needed for the development and deployment of strategic-weapon systems, our decisions to procure such systems are necessarily predicated upon predictions of some other party's capabilities ten years hence.

It is the incestuous relationship between these three factors which generates and drives on the strategic arms race and which also makes it so much harder for us to analyze and understand the operation of deterrence itself. For that reason, the one task which seems to me to be as urgent as it is difficult is to bring our techniques for the assessment of intention more closely into line with the efficiency of our techniques for the judgment of capability. That is why, with great respect to this audience, my own hopes of containing the strategic arms race, of consolidating deterrence, and of promoting a recognition of mutual interest between major powers lie not so much with natural scientists, who can both increase and interpret capability, as with social scientists and politicians, who may be able to help us to understand better the complex of psychological, sociological, and political pressures which generate strategic intention at the international level.

We must, as I see it, secure a much closer cooperation between natural scientists, social scientists, and politicians for the purposes which we all have in mind. The task of the social scientist and politician is obvious enough: to establish a better understanding of each other's political and social systems and of the decisions and policies which are likely to emerge from them, and thereafter to implement that understanding in political ways. The task of the natural scientists is more varied. One aspect to which I would call particular attention is, however, that of technical means of controlling agreements to limit armament. This is a peculiarly complex problem. Since the end of World War II, all the arms-control agreements which we have attempted to handle or have concluded have, in essence, been agreements to prohibit. In other words, they have been agreements to ban some type of weapon or some form of activity completely. The control of such agreements is relatively easy, given the high visibility of any activity above a zero level. We now, however, confront an urgent need to negotiate agreements to limit armament, and especially strategic armament, at some level greater than zero. The control of agreements such as these is a great deal more difficult, for obvious reasons. I am not, however, convinced that it is impossible, and I would hope that it would be natural scientists who would establish its possibility and would develop better means for controlling and verifying the observance of agreements to limit strategic armament, both qualitatively and quantitatively, at levels which are substantially greater than the zero level to which all previous arms control agreements have been attached.

Summary of Discussion

The Structuring of Forces. Despite the real stability of the present strategic balance, we must not overlook the possibility that the introduction of new technology will lead the strategic force-structures into a crisis-unstable configuration. There is today no theory of crisis stability because there has been only one real crisis involving the possibility of nuclear exchange between the superpowers in the years since 1945. This fact, however, does not allow us to assume there will never be another. Although we can only guess what sorts of forces would prevent a nuclear exchange in future crises, it is important for the nuclear powers to configure their forces ahead of time in a way which will minimize the likelihood of escalation. It is just in this sense that a MIRVed force is destabilizing. Although the deployment of a MIRV system does not portend a first strike in cold blood, it may encourage leaders, in time of crisis, to think that by striking first they could actually save themselves an enormous amount of damage. Such reasoning is particularly dangerous if the control of offensive nuclear weapons is being delegated to relatively low echelons.

There is a distinct advantage to having a force structure which guarantees very high potential damage with very high confidence. Such a posture greatly reduces the self-doubt and the anxiety which derive from worst-case analysis and tends to be less sensitive to technological innovation. However, high damage with high confidence implies that the potential level of damage, if deterrence should fail, is raised.

Capability and Intentions. There is one sense in which reliance on assessment of capability is preferable and safer than reliance on assessment of intention. The time scale for changes in a nation's capability is much longer than for its intentions, especially in time of crisis. However, actions are always constrained by capability, irrespective of intention. Hence capability is the more reliable, if often more conservative, measure. One disturbing aspect of making judgments based on capability only is the tendency to generalize from the very limited area of strategic-weapons policy to the worldwide political intentions and the personality of a government.

Third-country Deterrence. The question was raised whether a country with a relatively smaller force can exist in a relationship of mutual deterrence with the superpowers, now that the level of forces of the two superpowers is so very high. In answer to this question, it was pointed out that the important variable may be what is the prize involved between the two countries. In the event of all-out war between the two superpowers, the prize to the victor (if victory were possible) would essentially be world hegemony. On the other hand, the prize between a smaller power and a superpower would be much less than that. Hence a superpower would be likely to take correspondingly fewer risks and would be willing to sustain correspondingly less damage in a war with a smaller power. For this reason, a smaller power may indeed be considered to be able to maintain a credible deterrent with respect to a superpower.

Further Complexities. The real world is really more complicated than the discussion has for the most part assumed. The reality of chemical and biological weapons which are well within the reach of small powers is rarely considered in strategic discussions. This may be because the existence of the Geneva Protocol precludes the adoption of strategic doctrines involving their use, or because the use of chemical and biological weapons by or against the superpowers does not introduce any new component since they already possess overwhelming nuclear deterrents. However there is a real possibility that these weapons could be used in so-called "local conflicts," which could then escalate to include the superpowers. The other aspect not considered is the fact that the strategic balance is now, or will soon be, three-sided rather than two-sided, as it was in the 1960s. The complications arising from the growing Chinese nuclear capability have not been fully explored.

Treaty Implications. The realization that the present balance is rather stable and that very much lower levels of capability would be sufficient to maintain mutual deterrence has major implications as far as a treaty is concerned. The two superpowers can very comfortably, by running almost no risk, enter into an agreement relying only on unilateral verification. It is very important that the present negotiations be viewed in this framework. To avoid getting caught in the kind of entanglements that destroyed the total test ban treaty, it must be widely recognized that today only relatively major changes would make any difference in the strategic balance. The major enforcement mechanism of an arms-control agreement between the superpowers is the possibility that one side would bear the responsibility for the breakdown of the treaty following a violation. Since the political penalties of this responsibility are very great, each side has a major incentive not to violate the treaty.

Given this situation, the case was made that an arms-control treaty with a rather loose, informal, and subjective verification system is more stable than one whose enforcement is formal and objective. With the latter type, possible violations would be continually coming to light in public, forcing both sides to decide whether or not to abrogate the treaty. With informal and subjective verification, public pressure for abrogation need never build up.

On the Question of the Development of Military Technology

V. Emelyanov

Professor Valentine was quite right in saying that whenever a new discovery is made in science, the devil is quick to capture it while angels start a lengthy discussion on how to use the discovery in the most efficient way.

At the dawn of civilization the primate ancestors of modern human beings, as soon as they adopted the vertical state and freed their arms for moving, thus turning into two-legged creatures, the first thing they did was to take a weapon—a club—into their free hands. Then man started to improve his weapons, which became more and more efficient with the progress of human knowledge in science and technology.

In his book *The Dawn Warriors* (Atlantic Monthly Press, Boston, 1969), Professor Robert Bigelow, of Canterbury University (New Zealand), notes with regret that a scientific definition of man should perhaps be "a creature causing wars" rather than "a creature making tools." The whole history of civilization is full of wars and nowadays, when a number of brilliant discoveries in science have been made, a broad range of research is carried out with the aim of creating new means of mass annihilation of man.

Vasily S. Emelyanov, recently retired as longtime Soviet Representative to the International Atomic Energy Agency of the U.N., is now Chairman of the Disarmament Commission of the U.S.S.R. Academy of Sciences.

The participation of men of science in manufacturing means of military technology is ever growing. This started long ago and at all stages of human development scientists have been involved in creating armament.

Suffice it to recall that the greatest mathematician of the ancient world, Archimedes, dealt not only with mathematical calculations but was engaged in designing fortifications and many other military devices, including a gigantic catapult. Leonardo da Vinci, who was famous for being an artist, a scientist, and a prominent engineer, took part in developing many kinds of armament. In a letter to Sforza, the ruler of the principality of Milan, da Vinci offered to create any instrument which might be of use to the government: "military bridges, mortars, mines, military chariots, catapults and other machines of marvellous efficiency not in common use" (S. Zuckerman, *Scientists and War*, pp. 4–5).

Michelangelo, who was world famous as an artist and sculptor, was for a time the chief engineer in charge of erecting fortifications in Florence.

In the middle of the sixteenth century, when artillery began to appear, mathematicians were invited to take part in military research.

Nicolo Tartillia, an Italian mathematician, devoted himself not only to problems of pure mathematics but was also engaged in applied subjects as well: mechanics, ballistics, geodesy. In his works, Tartillia displayed calculations of the trajectories of moving shells. It was Tartillia who established the dependence between the shell's weight and its caliber, thus taking the first step in ballistics.

One of the founders of modern exact natural science, Galileo Galilei, by establishing the laws of force, proved that a shell would follow a parabolic trajectory.

One can make a long list of prominent scientists, engineers, and inventors who have participated in creating various kinds of weapons and military equipment. The whole history of metal production is, for instance, connected with the production of arms; and many scientists, engineers, and inventors in the field of metallurgy started their activities by producing metals for armament.

In the eighteenth century, classical science was completely developed. The main laws discovered by Galileo, Descartes, Newton and other prominent thinkers of the sixteenth and seventeenth centuries were considered as established, while Newton's laws of mechanics were regarded as the basis of the whole edifice of science

At that time people "believed in mechanical explanations of nature" and considered physics as a mechanics of a more complicated variety, calling it molecular mechanics. "Argument arose on the question of replacing physics by mechanics, and the details of mechanisms": that was the opinion of Abel Ray in his book *The Theory of the Physics of Contemporary Physicists*.

The industrial revolution of the eighteenth century gave impetus to tremendous technical progress that made the utmost use of the achievements in classical science.

Science opened up new laws and regularities in chemistry, the structure of molecules, the transformation of energy. The classical application of energy, which had solar radiation as its basis, gave impetus to the development of industrial enterprises.

Organic fuel was used for melting metals and the creation of machine techniques.

The source of energy has undergone considerable change: first it was the energy of falling water and wind; then it was the energy produced by organic fuels—logs of wood, coal, oil, natural gases. But the essence, in fact, didn't change at all; it was energy whose origin was closely connected with the sun. The sun gives birth to all classical sources of energy.

The nineteenth century was an era of experimental science in which numerous investigations were made and brought to completion. But at the forefront of classical science there emerged new ideas. The end of the eighteenth century and the whole period of the nineteenth century were famous for the discoveries that laid the foundations of modern science and technology and, at the same time, it was a period when the basis was laid for the destruction of a number of basic concepts. Experimental science, which received a remarkable impetus at the end of the nineteenth century, revealed new laws of nature, while technology, which was gaining strength, created favorable conditions for scientific experiments by providing the necessary instruments, devices, and materials and by making use of new discoveries in science; the ever-expanding horizons of knowledge accelerated the development of industry.

Production was becoming an "experimental science with strongly evident features of material—technical characteristics"—this was the conclusion of Karl Marx, who closely observed and evaluated the achievements of science and its impact on industry. Especially outstanding were the successes in physics and chemistry, the sciences thanks to which revolutionary changes are under way now.

But none of the discoveries in science and the breakthroughs in technology caused any fear or anxiety about the future. Never in the history of civilization did man feel a concern for the future of mankind, for the fate of all of us and of those dear to us, as he does now; there can be no comparison. Science has opened up ways of releasing mighty forces of nature such as are unprecedented in history.

The atom problem and rocketry—these are the main forces of the scientific-technological revolution now under way in the world. They brought to life new fields of science and branches of technology and led to a fresh approach to agelong scientific concepts. As these two problems were dealt with, new branches of industry and new technological processes came into existence, and the explosions of atomic bombs were followed by intensive activities of scientists whose ideas and thoughts, once in a dormant state, were now on the path of practical use. Extremely short became the periods for the realization of a number of scientific discoveries and research work. Thus, if the gap between Galvani's discovery of the phenomenon of electrical current and a first electric station was a century wide, it took scientists and engineers only three years to embark upon the program of the first atomic station; three years also elapsed between the first experiment on electron concentration within a ruby crystal and the first laser device. All this provides a vivid example of the scale of the scientific-technical revolution, which opens up enormous possibilities for practical employment of scientific discoveries for the benefit of mankind. But, at the same time, it gives rise to great concern for man's future, since military experts as well are involved in these turbulent activities. The stock of nuclear arms is sufficient to bring unprecedented disaster upon the world; yet the stockpiling continues and, what is more, means of delivery are constantly perfected; their efficiency is increased and possibilities are developed for providing guidance adequate to destroy their targets. A report prepared by representatives of twelve countries, at the request of the General Assembly of the United Nations, shows vividly and impartially what the consequences might be if nuclear arms were to be used in case of a third World War.

Robert Bigelow, in his book, notes that if formerly people used to rally and unite to take part in a conflict and fight a battle, now there is an urgent necessity of rallying once again, but this time with the aim of not allowing the destruction of what has been created by human civilization and the burying under its ruins hundreds of millions of people. This rallying is necessary, in Bigelow's opinion, if we are to survive.

People all over the world are more and more aware of the looming menace, but so great is the speed at which the danger of cataclysm is growing that really vigorous efforts are necessary to avert this danger to mankind.

At present many people understand that the results of the scientific-technical revolution now under way can be successfully used for solving the tasks once set by the most brilliant minds of humanity. In the *Bulletin of the Atomic Scientists* (February 1970) Eugene Rabinowitch writes that, from the economic point of view, war is no longer necessary, particularly now that its destructive force makes it politically irrational and morally unacceptable. At the same time, the threat of organic fuel running low, which was looming as a heavy cloud over highly developed countries, has been dispersed now that nuclear energy has come into existence; the horrible phantom of mass hunger, which was so positively predicted just a few years ago for the next decade, has now been eliminated, thanks to the "green revolution" which will increase twofold and threefold the grain yield in countries that badly need it; these are the thoughts of Dr. Rabinowitch in the *Bulletin*.

The peculiarity of our present stage is that the world now is in a state of imbalance, which means that an insignificant effort will be enough to change the balance. This, unfortunately, is beyond the understanding of many statesmen, but scientists take it for granted. This goes to prove that it is necessary for scientists to be responsible. It is necessary to put an end to the ominous escalation of the manufacturing of means of mass extermination of people. The present talks on disarmament, successful as they may be, are still inadequate and the negotiations themselves do not take into account the development of new types of armaments. No sooner had the United Nations finished discussing the question of the dangers of nuclear war, when the members started discussing the question of a new danger—this time the danger of chemical and bacteriological weapons.

The talks on disarmament are unfavorably affected by the current state of world affairs. When a bomb is exploded, no one can hear the flute—this is a well-known fact. But it is not only war that prevents the search for paths to peace; propaganda for war leads to an extremely unfavorable atmosphere, with its seeds of hostility and distrust. Even the debate itself on questions on armaments brings about unfavorable conditions and encourages further escalation in new research and the manufacturing of nuclear arms.

War propaganda and calls for aggression are formal pretexts for developing work on armaments and block the talks on disarmament.

Aggression and war propaganda are alien to the Soviet people. From the very first days of its existence, the Soviet State has fought for peace and establishment of neighborly relations with all states and for the adoption by all states of a policy of coexistence of states with different social orders. In our country, war propaganda is prohibited by law. In our country there are no people or groups of people whose welfare and prosperity depend on the manufacture of armaments, on the activities of war industries. That is why our government can so easily make such decisions—it is not under the pressure of those who are keenly interested in manufacturing armaments.

At the same time, in a number of countries of the capitalist world, especially in highly developed countries, there have emerged powerful groups of manufacture of arms that are particularly interested in their production.

The problem of preserving peace is a common matter of all people inhabiting our planet. The minds of the people, who are anxious about the fate of the world, will find a solution which will avert the approaching disaster.

Local Conflicts: Comments by R. Ramana

At this meeting most of the discussions have been on the arms race that exists between the superpowers, involving the use of ICBMs and anti-ballistic-missile systems. For a person who is not fully conversant with the reasons for this arms race, it seems that there exist no valid reasons why the two superpowers, which are both prosperous and big, should want to enter into an arms race purely for its own sake. One case in which the superpowers might be forced to use weapons of destruction of this magnitude seems to be when they get involved in smaller conflicts in various parts of the world which then escalate into a large-scale war. One would like to examine how the new technologies can control local conflicts and how they can help to stalemate the situation within a short time.

The development of ICBMs and the anti-ballistic-missile systems are of no relevance in such local conflicts, as these are usually concerned with border problems. In any case, the cost of such weapons of destruction prohibit their use in small conflicts. The recent developments in surveillance, intelligence, etc. are of some significance from the point of view that information about any possibility of local conflicts can be obtained sufficiently early for necessary precautions to be taken before a major conflict develops. Radar and infrared methods are already in use, but because local conflicts usually involve guerilla warfare, the best intelligence is still obtained by espionage methods. One has to study whether the new developments can in fact play a part in the prediction of movements of guerilla forces. It was very interesting to hear that by means of sonar methods, it is likely that the sea will become "transparent," i.e., one will be able to track a submarine down at any time in any place in the sea. This is of the greatest significance to countries which have long coastlines. The existence of a potential enemy's submarine close to the border is always an irritation and may lead to local and later to widespread conflicts.

The most important new technology that is of significance in the area of local conflicts is the use of nuclear weapons. I have been asked very often at this conference and elsewhere whether India is going to produce atomic bombs. The reason for such questions being asked is that India has been operating a natural uranium reactor for some years, together with a reprocessing plant and, as Dr. Mark stated at this meeting, it is possible to use reactor-grade plutonium for producing nuclear explosions. The situation has become more complicated because of the fact that India has opposed the present safeguard methods of the International Atomic Energy Agency and the nonproliferation treaty. Whatever be the tactical benefits of having a nuclear weapon—either for actual deployment or for the purpose of prestige—India has decided not to make nuclear weapons, purely for the reason that it is not good to produce such weapons. It feels that the nonproliferation treaty, as it stands, is very much one-sided and that the proposed system of safeguards will not safeguard anything. Rather, they may lead to a wide-scale racket in plutonium, much the same way as there exists large-scale involvement of governments in the smuggling of arms. India therefore feels that disarmament with respect to nuclear weapons can only start by means of examples from the superpowers. However, from all present indications, nuclear weapon testing is in full progress in both these countries.

Summary of Discussion on the Political Implications

Controlling the Arms Race. If a tradition could be established by which both superpowers released in advance an accurate and reliable account of their intentions in the strategic realm, many of the uncertainties which feed the arms race could be eliminated. If, over a period of time, such announcements were to become reliable enough that both sides could come to trust them, then the disparity between the possible and the likely would be diminished.

It may be, on the other hand, that such a simple and obvious approach to the problem is not possible. Some participants felt that it was unimaginable that the military would give up their claim to secrecy. Rather, an indirect approach should be adopted, aimed at changing attitudes over a wide sector of public life in the opposing countries, until it becomes much more generally accepted that it is in everybody's long-term interest to encourage wider international collaboration. This would eventually bring changes in the political climate at the top levels of both countries. Furthermore, there are many other problems of a global nature which mankind will soon have to face. If international cooperation could be arranged for dealing with these issues, then the overall political climate may begin to change.

Verification of an Arms-control Agreement. In the discussions there emerged two apparently extreme positions on the employment of technology to assist in the verification of an arms-control treaty between the superpowers. On the one hand, it was thought that recognition of the overall stability of the strategic balance and its insensitivity to small changes permit independent, unilateral verification by both sides. Furthermore, it was argued, the treaty could be written in general terms, without getting bogged down in minor technical details. On the other hand, it was suggested that cooperative reconnaissance, employing "transparent black boxes" (more accurately called "remote unmanned sensors"), should be employed to verify compliance with a treaty. These boxes would employ the latest sensor technology to transmit limited and predetermined information sufficient to give the other side a high degree of confidence that the treaty was not being violated. It was recognized that the employment of transparent black boxes could not be discussed adequately in the abstract and that they were better treated case by case, with particular, limited objectives in mind.

These two points of view are not necessarily contradictory. It is possible that different types of agreements could be written based on the two different approaches to verification. Also, the possibility exists of employing remote unmanned sensors in conjunction with other, perhaps unilateral, means of verification.

Hopes for SALT. There was a considerable degree of agreement on the desirability of reaching a consensus at the SALT negotiations on the following major points:

1. *Prohibition of a system of sonar arrays in the oceans.* Such a provision would both assure that an ocean-wide ASW system based on sonar would never be deployed and alleviate the fears about the vulnerability of sea-based forces. Because the deployment of such a system would involve a vast logistical undertaking, which could not be done in a clandestine manner, this provision could be easily monitored.

2. *Designation of large areas of the ocean where one or the other of the superpowers would agree not to place any of its military systems but which would be available to the other.* Such an agreement would eliminate the concern about hunter-killer submarines. Furthermore, if these areas were adjacent to the coastlines of each nation, the fear of a large massing of submarines off the coast for a preemptive attack would disappear.

These two provisions would allow each country to have a high degree of confidence in the continuing reliability of its sea-based deterrents. This confidence could then permit agreement on the limitation of land-based systems.

3. *Gradual reduction in the number of land-based missiles.*

4. *A very strict regulation of the deployment of ABM systems.* With the reduction in reliance on land-based systems, hard-point defense would become unnecessary. In order to guarantee the deterrent capability of each country, no appreciable city-defense ABM system should be permitted. There is no compelling reason to maintain ABM systems around Moscow and Washington only. However, if each

side were to maintain appreciable force levels and penetration aids, small ABM systems would not be very disturbing.

Restriction on the number of submarines would be feasible, but this does not seem to be important. In general, it would be very difficult to attempt to control the kinds of warheads and penetration aids deployed on missiles.

It was felt that the United States could design around the problem of the residual ABM capability of SAM installations.

The existence of a Chinese strategic capability might perhaps complicate the possibilities of incorporating these points in a treaty. It has often been suggested that both the United States and the Soviet Union should proceed with ABM and other systems, thus maintaining their own balance while, at the same time, rendering totally ineffective the much smaller strategic capability of China. However, a completely reliable anti-Chinese ABM system or even an air-defense system appears to be impossible from a technical point of view. One can not have the type of an umbrella which would deny China any possibility at all of delivering nuclear weapons. However, two administrations in the United States have pursued this line of thought and have decided to deploy anti-Chinese ABM systems. The present administration has argued that the Soviet Union would certainly not want to give up the possibility of an anti-Chinese ABM system and that therefore little agreement was possible on this question. If the indications from Vienna are correct, it seems that this estimate of Soviet intentions is entirely incorrect and that the strong regulation of ABM systems is indeed possible if the present U.S. administration would agree to it.

Other Possible Areas of Agreement. Aside from the possibility of limiting strategic weapons through SALT, other areas of agreement have been or could be possible. These include the demilitarization of the Antarctic, outer space, and the ocean floor. These agreements are important because they prevent the expansion of the arms race into new geographical areas. They also tend to impose restrictions on the associated military research and development. Whereas, in the past, civilian technology has often been greatly or totally derived from military technology, in those fields where agreement has been reached, the opposite may become the case. This would mean that the amount of military R&D which would actually be performed in these areas would be orders of magnitude less than it would have been without agreement.

The Future Role of Pugwash. The suggestion was made that, in the future, such Pugwash meetings of scientists might have three main objectives. First, there is the very necessary task of continuing to understand the current military situation to the extent possible from unclassified information.

Second, there is an equally important educational role: setting the issues in their true context and interpreting the systems-analysis jargon of defense planners so that laymen can understand and participate in the debates on these questions. One example is the oft-quoted phrase "damage limitation." Scientists should explain that there will always be some (not necessarily small) chance that one or more megaton weapons will slip through any defense system, even if cities are defended; that this would create a calamity on a scale which would dwarf Hiroshima or Dresden; and that, because of the complexity of modern society, at least three times as many people would be killed through hunger, disease, or exposure as were killed or seriously injured by the initial attack. This may not be a convincing argument against ABM deployment—which of course raises many other complex issues—but it is at least a factor which deserves to be weighed carefully, and in public.

Third, there is the question of what, if anything, can be done to get around the military problem by helping to change political attitudes. This is essentially a long-term aim for which, in Pugwash, the main asset is a tradition of free exchange of views between professionals who respect each other's competence. It seems at least conceivable that the scope of the discussions could be broadened to include other problems of common international concern, preferably chosen to bring in a wider range of interests, and hence indirectly a larger slice of the whole governmental machinery. Since many of the world's problems arise from the side effects of the industrial revolution, scientists ought to be well placed to consider them.

Index